Smart Underground Space and Infrastructures - Lille (SUS-Lille 2019)

MATEC Web of Conferences Volume 295

Villeneuve d'Ascq, France
8 - 10 October 2019

Editors:

Isam Shahrour
Xiong-Yao Xie
Hanbing Bian

ISBN: 978-1-5108-9777-9

Printed from e-media with permission by:

Curran Associates, Inc.
57 Morehouse Lane
Red Hook, NY 12571

Some format issues inherent in the e-media version may also appear in this print version.

For additional information, please contact EDP Sciences – Web of Conferences at the address below.

EDP Sciences – Web of Conferences
17, Avenue du Hoggar
Parc d'Activité de Courtabœuf
BP 112
F-91944 Les Ulis Cedex A
France

Phone: +33 (0) 1 69 18 75 75
Fax: +33 (0) 1 69 28 84 91

contact-edps@webofconferences.org

Additional copies of this publication are available from:

Curran Associates, Inc.
57 Morehouse Lane
Red Hook, NY 12571 USA
Phone: 845-758-0400
Fax: 845-758-2633
Email: curran@proceedings.com
Web: www.proceedings.com

TABLE OF CONTENTS

A thermal control methodology based on a predictive model for indoor heating management

Makram Abdellatif, Julien Chamoin, Didier Defer

YNCREA HEI, Univ. Artois, IMT Lille Douai, Univ. Lille, EA 4515, Laboratoire de Génie Civil et géo-Environnement (LGCgE), F-59800 Lille, France

Makram.abdellatif@yncrea.fr

Abstract. This paper presents a thermal control methodology based on an indoor temperature predictive model that regulates heating systems. Future indoor temperature is predicted from historical data of indoor temperature and other outside parameters. A control strategy is defined afterwards in order to retrieve the optimal parameters. The building simulation shows that the developped methodology provides satisfactory results by avoiding overheating and underheating periods and thus improving the thermal comfort and the energy consumption.

1. Introduction

1.1. Context

In France, the building domain is the most energy consuming sector. It dominates the total consumed energy share with 44% (compared to 33% per example for the transport sector) and a CO2 emission that overpassed 123 million tonnes in 2015 (1).

In this context, France has established an energy policy for buildings with the aim of reducing both of the greenhouse gas emissions by a factor of four by 2050 and of at least 40% the energy consumption of government and public buildings by 2020 (1). This policy is called *Factor 4*. These commitments concern new and retrofited buildings. Consequently, in order to reach these objectives, energy efficiency improvement is a must. This has been regulated in France under the Energy Performance Contract (EPC) law which guarantees the improvement of the buildings energy efficiency based on several standards after an energy renovation operation (1).

Generally, in the building life cycle, a significant difference is observed between the energy consumption predicted during the design phase and the real consumed energy during the operating phase. This difference is often called Energy Performance Gap (EPGAP) (2).

This EPGAP is due to several factors:
- During the design phase: incorrect sizing and unsuitable simulation tools,
- During the construction phase: poor quality of equipment and materials,
- During the verification phase: lack of systems verification and work execution,
- During the operation phase: occupant behaviour, weather and system control.

This paper presents a control methodology for indoor heating systems that optimizes indoor temperature and energy consumption in real time. It is organized as follows: Section 1 introduces the context of this work. Section 2 describes the proposed methodology. Section 3 presents an application case based on a simulated building. Finally, Section 4 concludes the paper and expresses some wayforwards of this work.

1.2. Building's Thermal modeling

Usually before each construction or renovation operation, the building is modelled in order to evaluate its performance and compare it with the requirements of current regulations. In the research and industry field, there are different methods for modeling building to study its energy efficiency. These methods can be classified as follows:
-White box: studies physical phenomena and can estimate them at a given time and in a given space, it is performed through a Dynamic Thermal Simulation (DTS) software (e.g. TRNSYS, EnergyPlus, etc.). It requires a good knowledge of the building (materials, geometry, Ventilation and Air Conditioning (HVAC) description, control strategies, occupancy, location, etc.). It can provide very detailed results but they can be very time-consuming (3),
-Black box: a model based on mathematical and statistical methods (Machine Learning, Markov Chain, etc.). It may require more or less important training time depending on the method used and the calculation time is generally low (3),
-Grey box: solution combining the two previous models requiring less learning data and less knowledge of physical phenomena (3). One of the most famous grey box models is the RxCy model (4).

As part of this work, the following criteria are set up to choose the suitable prediction model:
-Low calculation time: the model will be used to predict the indoor temperature and will also be used to evaluate a high number of control strategies,
-Adaptable: in the building, the year is composed of a heating period, a cooling period and an inter-season, the model must be efficient during these 3 periods,
-Generalizable: the model must be easily deployable on any type of building if the measurements are available.

Based on these criteria, it was decided to test Artificial Neural Networks (ANN) and multiple linear regression models. Indeed, many studies (5)(6)(7)(8) demonstrate the ability of ANNs to predict thermal data in the building sector. Nevertheless, the ANNs has shown a problem of robustness. Consequently it was decided to use the linear regression model. A comparison of the two methods will be discussed.

2. Control methodology

The porposed heating control methodology is based on a model insuring the predicting of the indoor temperature in order to avoid discomfort problems. The steps of the control methodology are presented in Figure 1.

 https://doi.org/10.1051/matecconf/201929501001

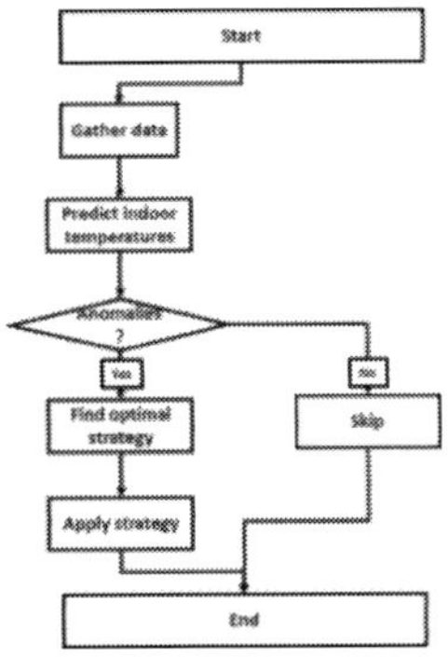

Figure 1: Proposed control heating methodology

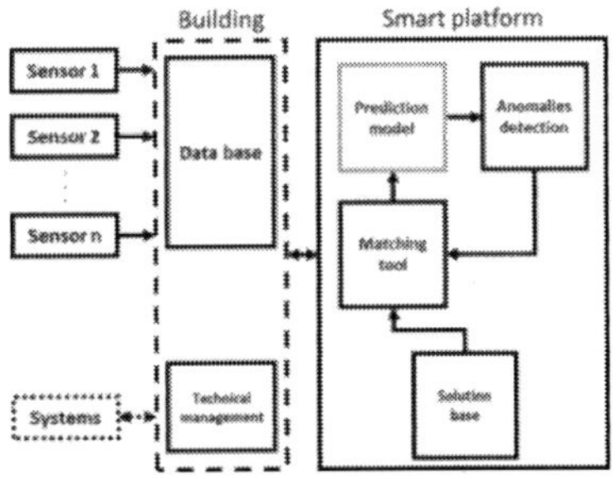

Figure 2: Functioning of the proposed platform

Based on this methodology, a platform was developed. The functioning of this platform is shown in Figure 2.

1.1. Step 1: Data gathering

In the first step, dynamic data (indoor measurements, HVAC and weather data) are gathered from sensors and Building Management System (BMS). Then collected data are structured and pre-processed (Normalization and Averaging).

1.2. Step 2: Predicting indoor temperature

A multiple linear regression model is implemented. To determine the model inputs, a correlation study was conducted. The most influencing inputs were selected. The data are normalized to be in the same scale. From the measured data, the averages of the last hour and the last two hours are calculated to take into account the inertia of the building. The final architecture (Figure 3) of the model is validated by varying the input parameters (adding or deleting one or more parameters). The prediction horizon is set to 3 hours which is sufficient to take into consideration the time response of the building while maintaining an acceptable model accuracy level (Figure 4). At each step, 12 data are predicted with a time step of 15 minutes.

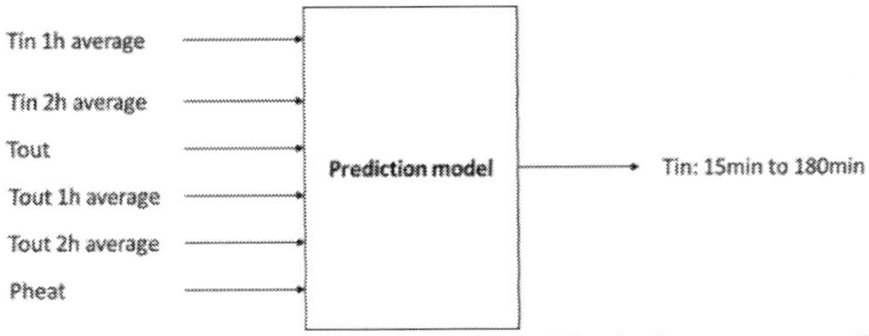

Figure 3: Prediction model of the indoor temperature

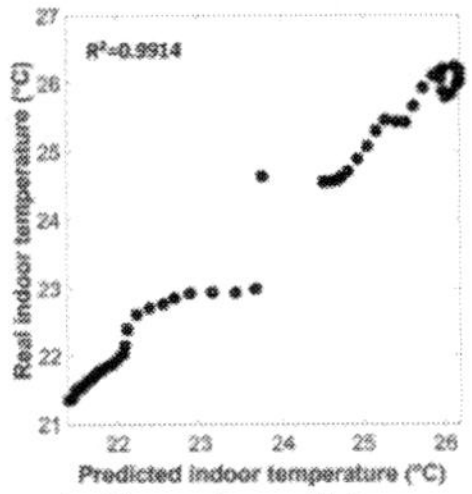

Figure 4: Results of the prediction model

1.3. Step 3: Detecting anomalies & finding optimal startegy

Using the predicted indoor temperature, the platform checks if the temperature is in the following comfort range during working hours (8am to 6pm): [18°C 25°C] (9) and during unworking hours: 16°C, while reducing the heating consumption as much as possible. An anomaly is reported if the temperature is outside this range.

1.4. Step 4: Application of the new strategy

The selected control strategy is tested with the prediction model and then sent to the BMS.

3. Application

Tests were carried out on a simulated building. This building is composed of two thermal zones (Figure 6) and equipped with a power-controlled heating system. Between 8am and 6pm, the power is set to 100% and 0% for the rest of the day. This type of control is close to reality and can ensure the comfort criteria defined above. The control curve of the heating system is shown in the Figure 5. The building is described in Table 1.

Table 1: Building description

	Area (m²)
Thermal zone 1	48.99
Thermal zone 2	11.86
Windows	North: 2 (1x2m), South: 1 (4x1.5), West: 1 (1x2m)
Door	East: 1 (1x2.5m)
Location	Lyon (France)

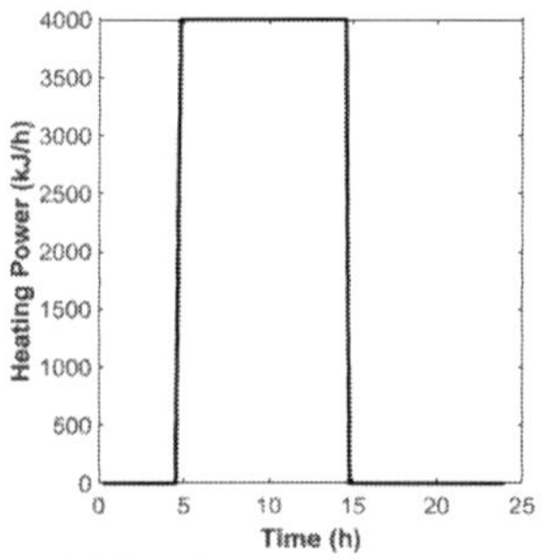

Figure 5: Heating setpoint power curve

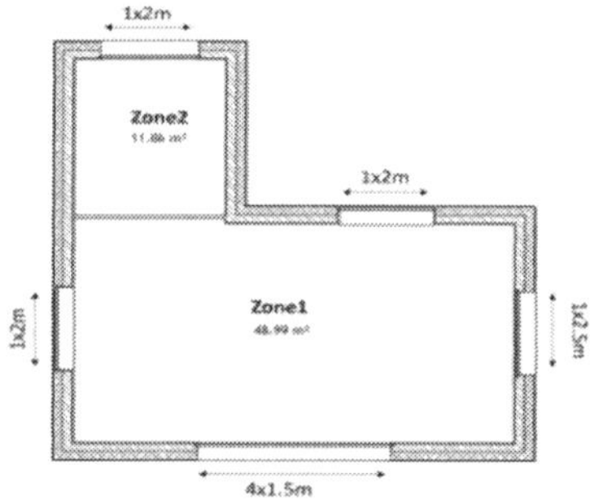

Figure 6: Plan of the fictious building

Based on previous historical data (from simulation results) the prediction model is trained then the indoor temperature is predicted. At every time step, the prediction model is retrained using previous data (150), this method is called *sliding-window* (7). The size of the sliding-window (training data size) was determined by performing tests. Figure 7 shows that the temperature exceeds 25°C which is the comfort limit defined. As a consequence, the data analysis detects anomalies. To face this anomaly, the best heating control strategy is defined for the next 3 hours. Since the prediction model is fast, all possible strategies are tested. For this applicatiuon case, the heating is controlled by an On/Off mode, then 4096 (corresponding to 2^{12} with 12: nombre of time steps to predict, 2: ON/OFF state) possibilities were tested. At each time step (15 minutes), heating is switched ON or OFF. Figure 7 compares real indoor temperature to predicted indoor temperature and Figure 8 shows the result obtained by applying the new heating control strategy from the 152th until the 155th hour, with data from the 3rd to the 151th hour are used to train the prediction model.

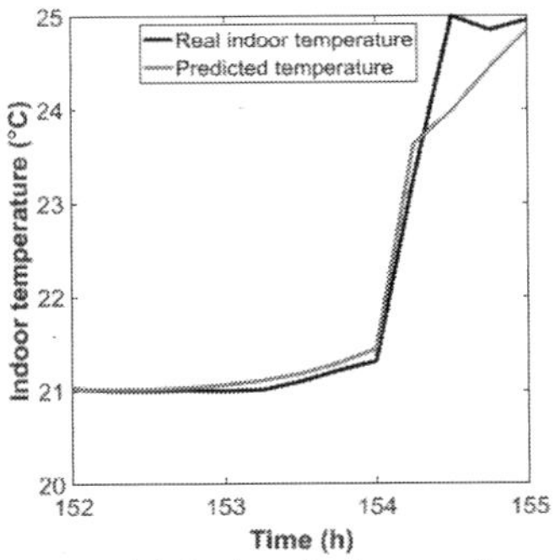

Figure 7: Initial simulation and prediction results

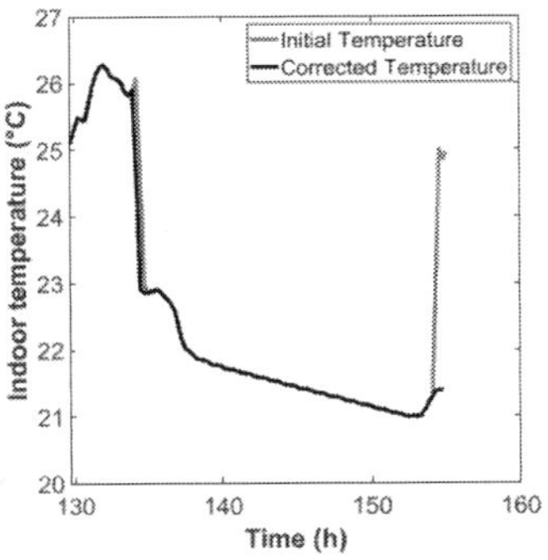

Figure 8: Corrected prediction results versus initial simulation results

This methodology is done in real time, obtaining the results shown in Figure 10. 30 hours of obtained result of the thermal zone 1 based on the simulated building is shown. Figure 9 shows the evolution of the indoor temperature and the heating power setpoints with the use of a standard control of the heating, while Figure 10 shows their evolution using the methodology proposed above. Comparing the data on the right of the two dotted lines representing the start time of the heating control correction, in Figure 10 the overheating zones are significantly reduced while respecting the comfort interval (here taken between 21 and 25°C for representativeness reasons) represented by the horizontal lines, which improves user's comfort and reduces energy consumption.

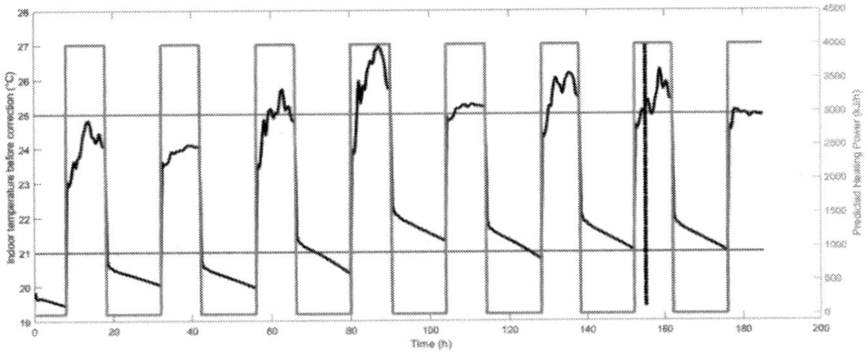

Figure 9: Simulation results with standrad heating control

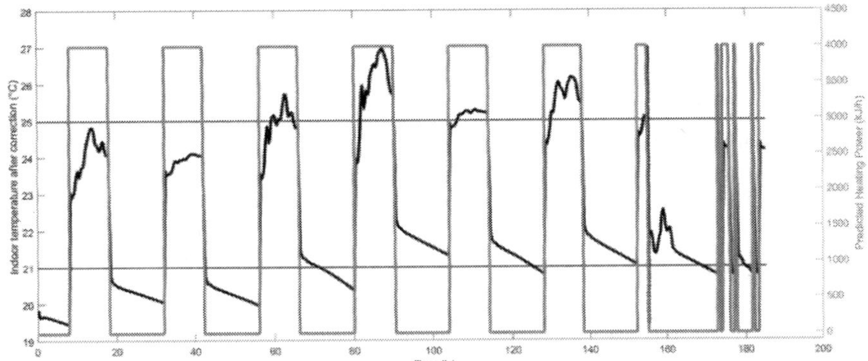

Figure 10: Simulation results with the proposed heating control methodology

4. Conclusion

This work aimed to develop thermal control methodology for reducing discomfort and energy consumption. In this context, a platform was developed based on a linear regression model to predict indoor temperature and to test possible control strategies. To validate this platform, a simulated building is studied using a BIM-based approach. Obtained results during heating period are promising. The heating control methodology allows the reduction of the heating energy consumption and improves the user comfort. In the next step, a real case-study will be examined. This study-case is a building located in Wasquehal (France) which has been equipped with a weather station and comfort sensors.

5. Acknowledgments

This research work is supported by *VINCI Construction France*, *Cegelec*, *Projex* and *Yncréa* as part of *Smart Building as nodes of Smart Grids* (SBnodesSG) industrial chair.

6. References

1. Énergie dans les bâtiments [Internet]. 2016. p. 1–9. Available from: https://www.ecologique-solidaire.gouv.fr/energie-dans-batiments
2. Kallab L, Chbeir R, Bourreau P, Brassier P, Mrissa M. HIT2GAP: Towards a better building energy management. Energy Procedia [Internet]. 2017;122:895–900. Available from: http://dx.doi.org/10.1016/j.egypro.2017.07.399
3. Foucquier A, Robert S, Suard F, Stéphan L, Jay A. State of the art in building modelling and energy performances prediction: A review. Renewable and Sustainable Energy Reviews. 2013;23(January 2018):272–88.
4. Andriamamonjy A, Klein R, Saelens D. Automated grey box model implementation using BIM and Modelica. Energy & Buildings [Internet]. 2019; Available from: https://doi.org/10.1016/j.enbuild.2019.01.046
5. Zouak. A comparison of linear and neural network ARX models applied to a prediction of the indoor temperature of a building. 2004;32–7.
6. Moon JW, Lee JH, Kim S. Evaluation of artificial neural network-based temperature control for optimum operation of building envelopes. Energies. 2014;7(11):7245–65.
7. Moon JW, Yoon SH, Kim S. Development of an artificial neural network model based thermal control logic for double skin envelopes in winter. Building and Environment [Internet]. 2013;61:149–59. Available from: http://dx.doi.org/10.1016/j.buildenv.2012.12.010
8. Moon JW, Kim JJ. ANN-based thermal control models for residential buildings. Building and Environment. 2010;45(7):1612–25.
9. Réglementation officielle , mesures de prévention , droit d ' alerte et de retrait [Internet]. 2018. Available from: http://cgtdgfip75.org/spip.php?article3846

An approach to design a distributed knowledge-based platform for energy efficient cities

Pavel Vitliemov[1], *Daniel Bratanov*[2], *Milko Marinov*[3]

1 Department of Management and Business Development, University of Ruse, Bulgaria

2 Department of Public Health and Social Activities, University of Ruse, Bulgaria

3 Department of Computer Systems&Technologies, University of Ruse, Bulgaria

Abstract. In this paper is presented an approach to design distributed knowledge-based software environment for energy efficient cities. All data gathered by the smart tools at domestic or public facilities can be accessed by the platform. Energy savings can be achieved as result from an optimized mix of both condition-based and preventive energy losses services that followed an online analysis of collected data. Algorithms and methods for an intelligent interpretation of the collected data can be developed within the platform in order to monitor the energy consumption process. Simulation of different scenarios forecasting the consumer's further behaviour in energy consumption will be the result of that intelligent interpretation.

1. Introduction

The Energy Efficiency Directive 2012/27/EU from 2012 to set a 20% energy savings target by 2020 (when compared to the projected use of energy in 2020) is almost equivalent to close 400 power stations. [1,2]

In December 2018, the revised Energy Efficiency Directive entered into force (amending Directive EU (2018/2002) updating some specific provisions and introducing some new elements. Above all, it establishes a headline EU energy efficiency target for 2030 of at least 32.5% (compared to projections) with a clause for a possible upwards revision by 2023. [3,4]

Households account for 25-30% of European carbon emissions [5], with 80%+ of household energy used on heating. Central to achieving energy conservation will be behaviour change in households, shifting towards more efficient domestic energy management [9]. This has been evidenced in a range of research on energy conservation [6,8]. Nowdays, it can be stated that a key barrier to household action on climate change is the sense that individuals cannot make a meaningful contribution to it, because is currently very difficult to measure the impact of an individual household, that make the sense that the issues are so large and is hard to personalize the impact an individual can have [10]. This makes the perspective for policy around Personal Carbon Allowances and emissions trading for households crucial from European Governments' perspective. [11]

The household action is currently difficult even if there was better information and can be stated that the household appliances and systems that use energy are rarely networked or automated intelligently. A major effort is required in order to making adjustments to settings to optimize environmental performance of different appliances and devices e.g. thermostat settings for heating, standby settings for appliances, temperature settings for washing machines and dryers etc.

It can be stated that holistic knowledge about energy efficiency potentials in cities is far from complete and nowdays a variety of individual strategies and approaches by different stakeholders are the basis in the strategic energy efficiency planning. An integrative approach to achieve the sustainable, energy efficient, smart city is a strategic issue by connecting scientific excellence and innovative enterprises in the energy sector. The pilot knowledge-based platforms can help all ambitious and well-organized cities to reduce energy use in Europe contributing to the EU's 20-20-20 targets ensuring following: [6,9,10]

- A real time Carbon footprint assessment, that depends on the sources of energy,
- A real time household energy efficiency rating that follows an analysis how the house is resulting in terms of heat loss and the efficiency of appliances,
- Recommendations for energy efficiency improvements based on the real time energy efficiency analysis,
- Recommendations for different usage profiles that match householders usage needs more energy efficiently,

- Recommendations for forward settings based on online information about developing weather patterns and past usage patterns during these times.

Technological, behavioural and structural factors are integrated as different dimensions in energy efficiency improvements as these capabilities are embedded in improved practices, e.g. of building and using houses in a sustainable manner. [3]

Energy smart cities start with behaviour change of both individuals and organisations. Understanding how people make choices on their energy consumption – whether unconsciously or consciously will help for designing energy saving policies that will rely on behaviour as very much context-driven and subject to a multitude of situational and structural drivers. [4]

The main objective of presented approach within the paper is to contribute to above challenges with developing suitable tools to enable municipalities to improve their local and global performance at energy efficiency. All these techiques, methods and technologies can be implemented if "smart tools", that represent public and resident infrastructures equipped with up-to-date sensor technology will deliver real-time data from the energy consumption process and the energy manager will be introduced to the energy consumption and be aware of the actual condition. A cost-effective and rapid creation of both knowledge- and technology-based energy management services is made possible.

The Web-based platform that coordinate all forms of forward and backwards collaboration will supporting any activities along the value-adding chain based on smart tools that delivers real time data from the energy consumption process. Based on forecasts, errors or energy loses can be interpreted in real time by the energy manager, or energy consumer, thus minimizing the time to respond and the correction time, according to the appropriate energy consumption level based on the equal environmental conditions.

2. Concept of proposed software platform architecture

The concept of the work is based on the distribution of a set of data collectors with build-in web services. This data collector will be called for short – Jini (fig.1) [12]. The Jini is an electronic diary. Each jini will be able to collect and manage data broadcasted from different energy measurement sensors. The jinies have the intelligence to analyze the incoming data and connect with the data management center in the cases where there is a tendency for change in the situation. Thanks to the Jini, the municipality and the energy supply companies are able to surveillance, record and manage the energy consumption in the resident and public area continuously for years. The actual values for each energy sours together with the information collected for the environmental conditions are saved in the data collector in order to be used later on for statistical analyses. This allows the owners to predict and overcome eventual malfunctions of the energy system and prevent energy losses. The Jini is self-controlled. It is capable to control its own load and working regimes.

Figure 1. A view of a Jini manufactured by Feller Engineering GmbH, Germany [11]

As shown on fig.2 a network of Jinies can be linked via the internet to the energy management system [8]. The Jinies can be placed in very different locations and feed with variety of data. An example of architecture of a Jini installed in resident house is given in fig.3.

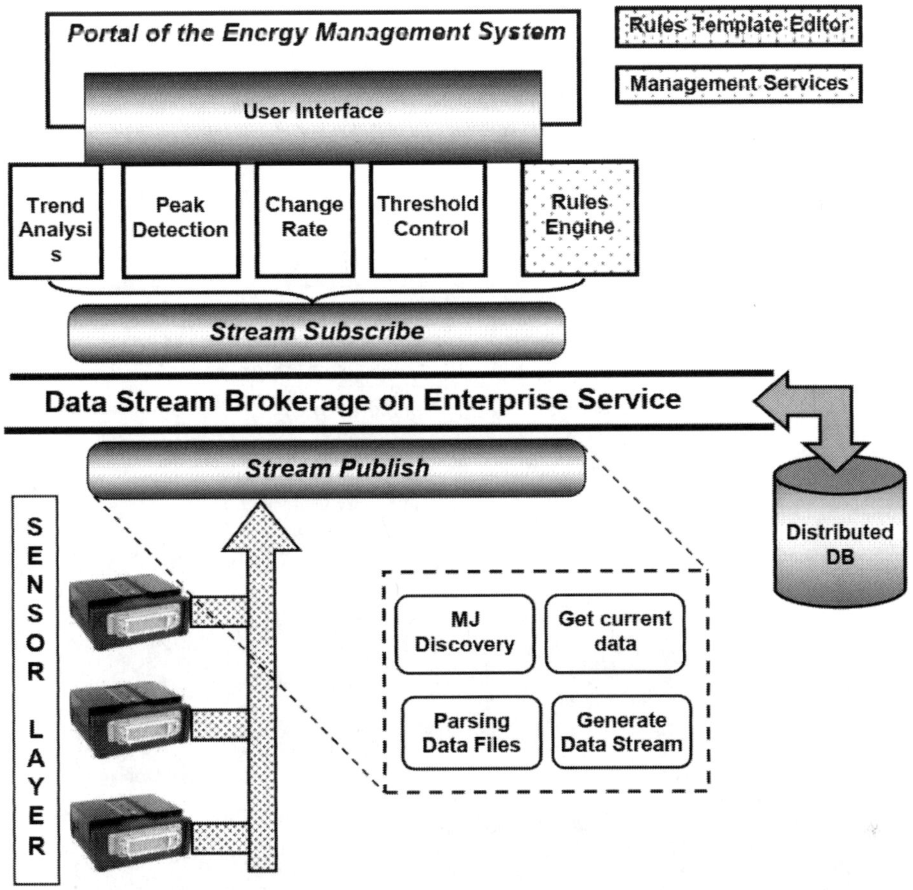

Figure 2. The architecture of Jini energy management software platform

The same architecture can be organized not only at domestic environment, but also to public buildings and even in the public transport. Many different systems are available on the market today that can measure and record the energy consumption. One of the basic disadvantages of those systems is in the lack of interpretation of the measured data. The idea that stands behind the architecture of Jini energy management software platform is that the collected data can be interpreted in different means by the ability of the end user to create different management profiles trough a rules engine. The rule engine provides the option for reconfiguration of the energy management system. This makes the energy management flexible and adaptable to the specific needs of the different users and energy plans available. A random setup of measurement units which data can be stored, analyzed and broadcasted by Jini is presented on fig. 4.

This feature provides a flexible way to access real time energy data gathered and an easy and flexible direct access to process data from any type of application for real time data processing. There is a data source (a publisher) that collects energy data and arrange them for a general provision, a broker that distributes the data on the bus, a subscriber that gets the data and put them within a XLM file to which an application can access. The ActiveMQ is used as broker activating the publish/subscribe services. ActiveMQ is an open source message broker that supports Java Message Service (JMS).

It is a standard Java API for creating, sending, receiving and reading messages. The communication between the broker and any application client is based on API (JMS interface). The logical and physical communication layer is represented from the Enterprise Service Bus component (ESB). ESB (ServiceMix/OpenESB) is a software component that lies between the energy applications and enables communication among them.

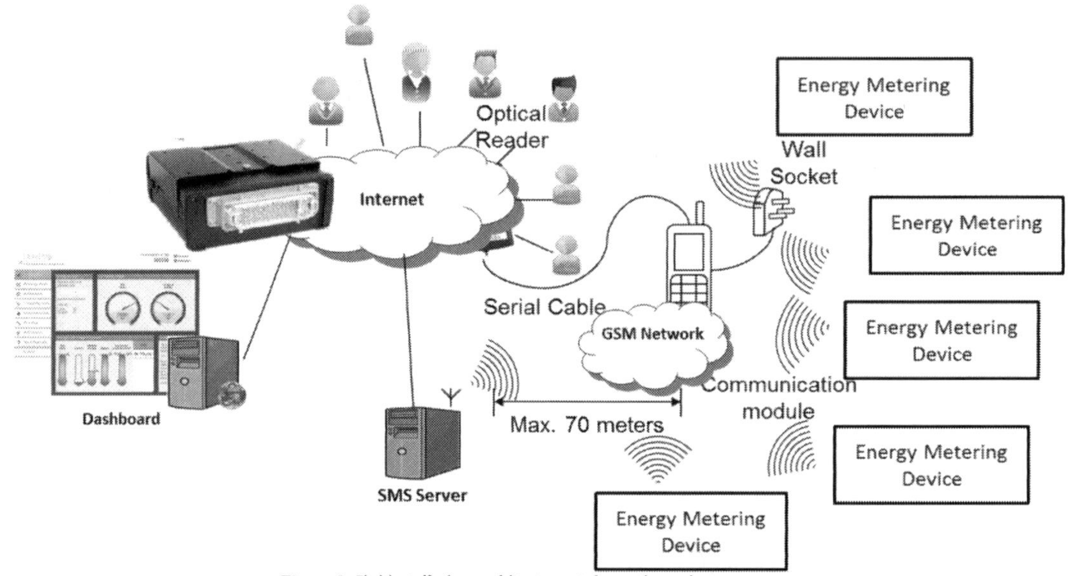

Figure 3. Jini installation architecture at domestic environment

Figure 4. Different sensor setting for measurement of energy consumption

The data collected by the Jini from the different sensors can be managed in two different methods. In the first method the data is transferred from the Jini to the Portal of the energy management system. An example of Graphical User Interface GUI of such portal is shown on fig.5, where the user has the ability to monitor the current energy consumption via the average energy consumption for the region.

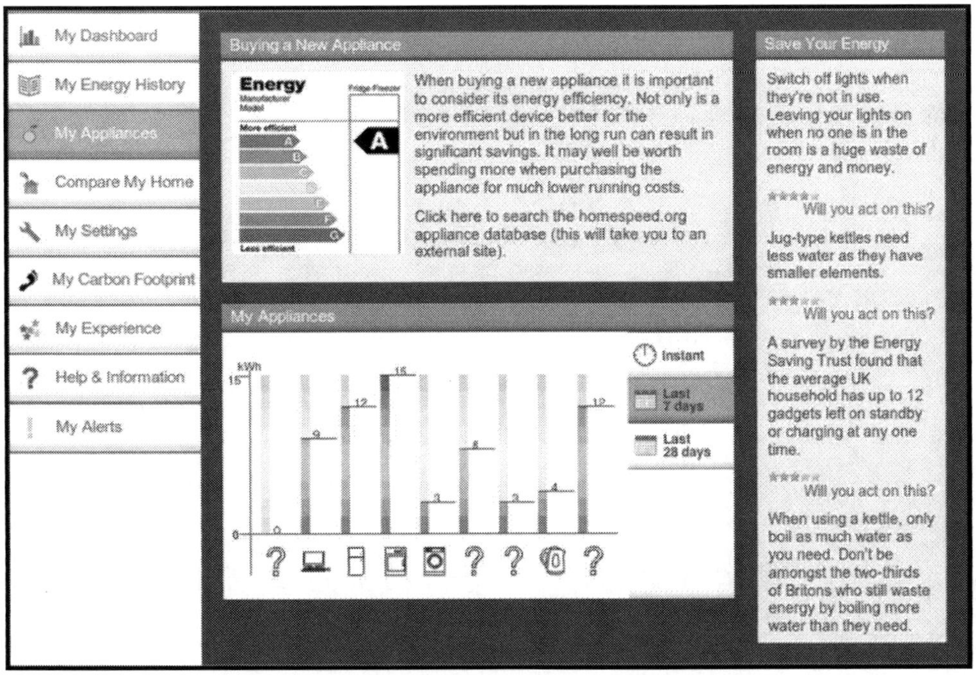

Figure 5. GUI of energy management portal

Each user of the portal has the ability to monitor his facilities. The monitoring process can be organized as overall monitoring or divided into monitoring of the different domestic appliances (fig.6) that are sending data to the Jini, respectively to the portal.

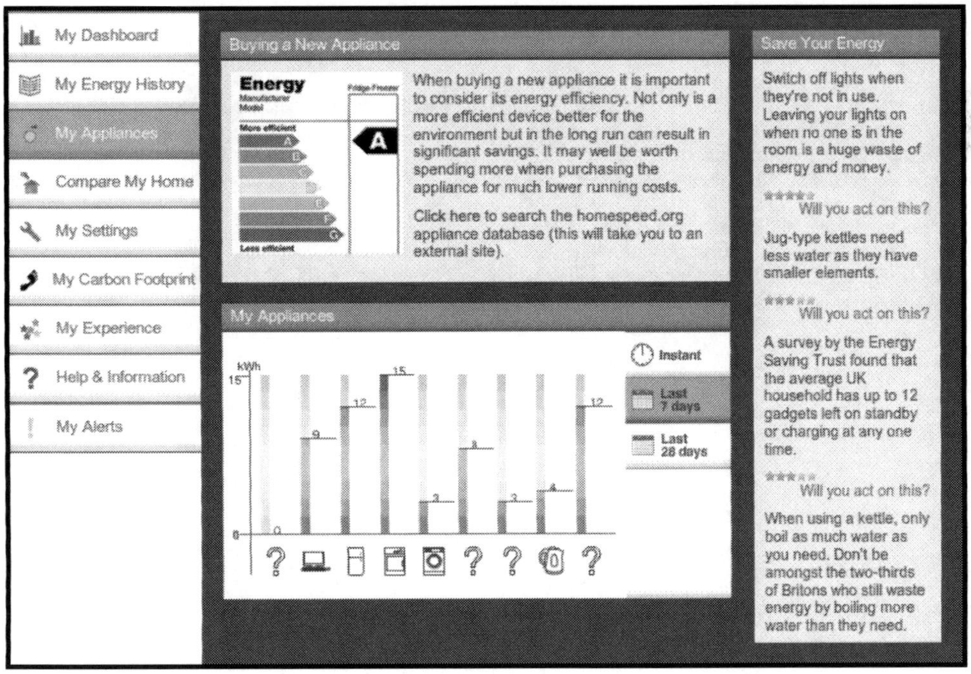

Figure 6. Labeling the domestic appliances to the management system

The data collected in the Portal can be visualized in variety of ways, but this process still cannot be considered as energy management as it provides only statistical data. In order to transfer the surveillance and monitoring system for energy consumption into an energy management system an integration environment must be developed. The data interpretation boxes can be considered as the library containing algorithms and methods that can be used within different workflows. Basically these tools are used within a process that could be entirely or partially supported from the proposed platform facilities.

3. Conclusions

An approach to design a distributed knowledge-based software environment for energy efficient cities that will gather the data collected by the smart tools, data loggers equipped with up-to-date sensor technology delivering real-time data from the energy consumption process at domestic and public facilities is proposed in the paper.

It can be concluded that the proposed platform can move beyond current smart metering energy 'input' models that monitor the levels of energy being used by a household, towards energy 'performance' models that bringing together sensor data in areas such as household heat loss and appliance performance and all the data from energy meter systems and sensors e.g. for gas and electricity, can be analysed against energy intelligence on target efficiency, average efficiency rates and environmental conditions in order to provide practical information to the householder in a 'dashboard' that can be accessed online.

References

1. Energy Efficiency Directive (2018) (amending Directive EU (2018/2002);
2. EU Commission (2014), [1] Climate Change and Major Projects, Available at: https://ec.europa.eu/clima/sites/clima/files/docs/major_projects_en.pdf;
3. EU Commission (2014); Commission Delegated Regulation (EU) N:480/2014, Available at: https://eur-lex.europa.eu/legal-content/EN/TXT/PDF/?uri=CELEX:32014R0480&from=en;
4. EU Commission (2016), EU Reference Scenario, Energy, transport and GHG emissions Trends to 2050; Available at: https://ec.europa.eu/energy/sites/ener/files/documents/ref2016_report_final-web.pdf;
5. Climate Change Programme Review, Sustainable Development Commission, UK, 2005;
6. Jackson, T., (2005) Motivating Sustainable Consumption. A Report to the Sustainable Development Research Network;
7. Milko Marinov, Pavel Vitliemov, Elitsa Popova. Towards Big Data and Internet of Things as Key Aspects of Energy Efficiency.// TEM Journal, 2017, No 6(3), pp. 427-435, ISSN 2217-8309.
8. Mobilising individual behaviour change through community initiatives, Centre for Sustainable Energy, Community Development Exchange, 2007;
9. Kirova M.; Pr. Velikova (2016); Risk management method for small photovoltaic plants, Book Series: Proceedings of the 10th International Conference on Business Excellence in Energy, Climate Change and Sustainability Location: Bucharest, Romania Date: March 03, 2016;
10. Jackson, T., (2005) Motivating Sustainable Consumption. A Report to the Sustainable Development Research Network;
11. A Rough Guide to Individual Carbon Trading, Simon Roberts and Joshua Thumim, Centre for Sustainable Energy, Department for Environment and Rural Affairs, 2006;
12. FELLER ENG (2018) Operator manual. Feller Engineering GmbH, Germany. Available at http://www.fellereng.de/, accessed December 2018.

Transformation of the water system of the Education City in Doha into a smart water system

Abdallah AL-NAEMI[1,2] *& Isam SHAHROUR*[1]
[1] *Laboratoire de Génie Civil et géo-Environnement, Université de Lille, 5900 Lille, France*
[2] *Qatar Foundation, Doha, Qatar*

Abstract

This paper presents the transformation of the water system of the Education City in Doha (Qatar) into a smart water system. This city covers an area of 14 km^2 and includes 80 buildings. The water system provides drinking, irrigation and fire protection services. It suffers from the use of fragmented management systems and from a lack of real-time monitoring, which result in a deterioration of efficiency of the water system and users' information The paper describes the water system and then the architecture of the smart water solution and its use for leak detection, water quality control and operation safety.

Key words: Smart Water, Leak, contamination, irrigation, fire-protection, Education City

1. Introduction

This paper concerns the transformation of the water system of the Education City in Doha (Qatar) into a smart water system. The Education City covers an area of 14 km^2 with 80 buildings and facilities. It was constructed by successive extensions of an existing area. As a result, the city uses fragmented tools for utilities management, which results in losses in efficiency and an increase in expenses.

Today, the modernization of water utilities becomes a "must" in order to reduce water losses, improve asset management, reduce water contamination risk and improve the quality of water services. According to several researches, the smart water concept constitutes an excellent solution to achieve this goal (Storey et al. 2011, Lambrou 2014, Farah & Shahrour 2017, Abou Rjeily et al. 2017, Mishra et al. 2018, Saab et al. 2018, Liu et al. 2019). This solution is based on a comprehensive real-time monitoring of the water utilities, share of data with concerned stakeholders and use of advanced tools such as Artificial Intelligence for data analysis and the transformation of collected data into operation data that ensure the optimal management of the water system.

This paper presents the transformation of the complex water system of the Education City into a smart water system. It presents successively the Education City, the water system and the transformation of this system into a smart system.

2. Presentation of the water system in the Education City

2.1. Presentation of the Education City

Figure 1 shows the development of the Education City in Doha. In 2002, it covered an area of 3 km^2, then it expanded to 6.1 km^2 in 2003 and to 7.5 km^2 in 2004. In 2016, it covered an area of 14 km^2. Figure 2 shows the master plan of the Education City. It is composed of 2 zones: North and South zones. Because of historical factors, the infrastructures of these two zones are independent. The Education City hosts around 80 buildings, which are used for (i) Higher Education (10 buildings), (ii) Pre-Education (3 buildings), (iii) Technology (9 buildings, (iv) Social services &

activities (14 buildings), (v) Culture, arts & heritage (5 buildings), (vi) Housing & hotel (6 buildings)and technical services (28 buildings).

2.2. Presentation of the water system

The water system provides drinking water, fire protection and irrigation services. Eight central plants are used as the main hub for water operation. Figure 3 shows the distribution of the central plants. Each zone of the Education City is served by four central plants. Figure 4 illustrates architecture of the water system. The central plant is directly supplied from the water company (Kahramaa). A water meter records the water supply of the central plant. Water is stored in a tank, which has a capacity to cover water needs for 2 days. Pumps are used for water transportation to the consumption sections. A utility tunnel hosts pipes for drinking water, irrigation and fire protection. Flow meters were installed to record buildings' water consumption. Each central plant uses a dedicated water management system.

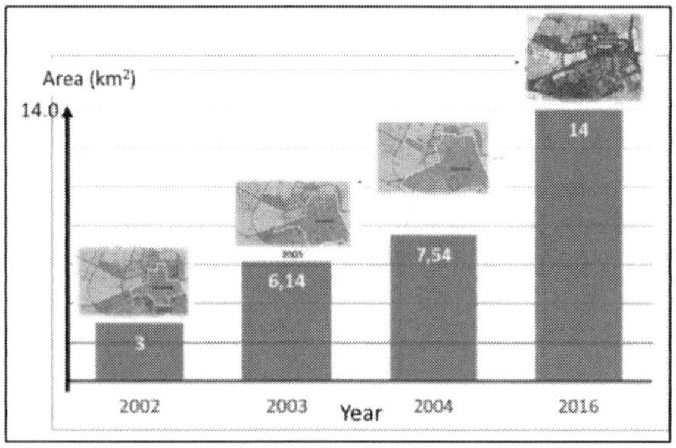

Figure 1: Stages of development of the Education City

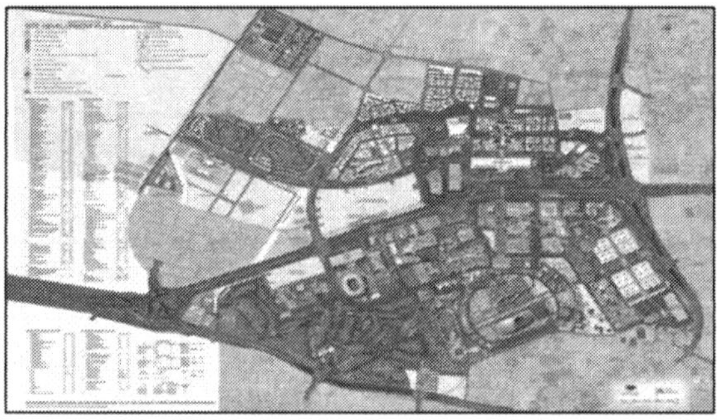

Figure 2. Master plan of the Education City

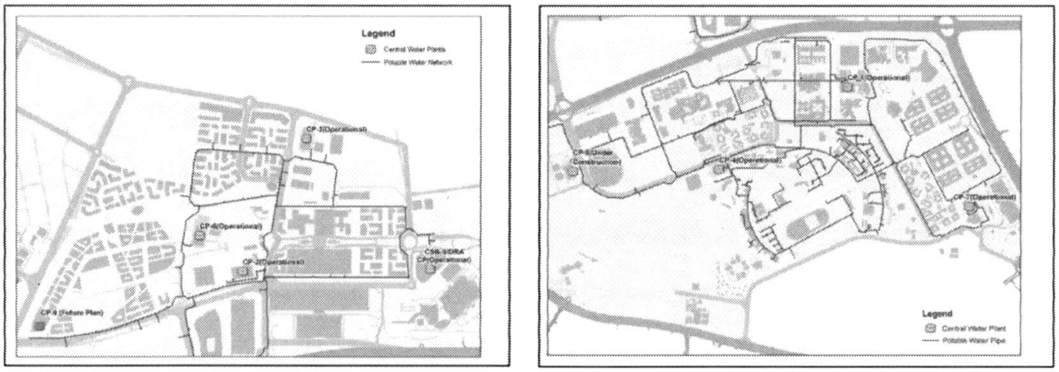

North zone South zone

Figure 2. Location of the central plants in the Education City

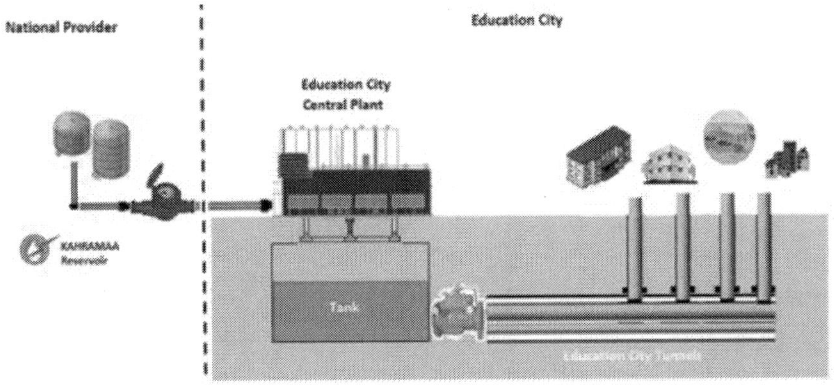

Figure 3: Organization of the water system of the Education City

3. Smart water solution

3.1. Solution architecture

The transformation of the water system of the Education City into a smart water system requires an integrated approach with a unique smart platform, which ensures data collection, data storage, data analysis, data visualization and operation control. The smart water system presented in this section is based on researches and achievements of SunRise project conducted at Lille University (Shahrour et al. 2017, Farah and Shahrour 2017, Saab et al. 2018 2018) as well as the outcome of the European project SmartWater4Europe (SW4EU, 2017).

The architecture of the smart water solution is presented in Figure 4. It includes a platform, which coordinates operations related to the smart management of the water system and an information system, which stores:

- Assets data of the water infrastructures, which are collected using GIS. All the components and attributes are integrated in the same system and could be visualized in GIS environment
- Data collected by the sensors, which concern the system security and optimal operation, such as water flow, water pressure and water quality.
- Maintenance and operation data.

Data analysis is conducted using engineering software such as those used in leak detection, contamination detection, operation control (pressure, flow, temperature,…). Artificial Intelligence is used for data analysis. It allows to learn from historical data, to identify operating trends and interaction as well as abnormal events and actions to be taken in case of an abnormal event. It allows also the optimal management of the water system.
Data visualization is conducted using GIS software as well as specific software to visualize the operating conditions, incidents and maintenance.
The results of data analysis are turned into operationl data (security and optimization) and transmitted to the water system through command equipment such as electronic valves, switches, motors,…

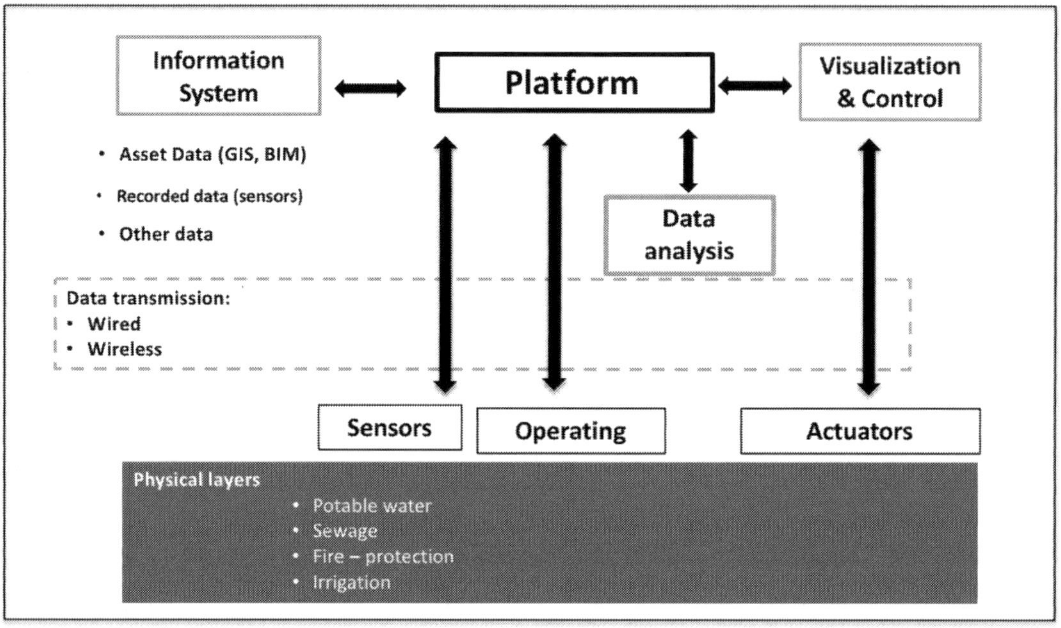

Figure 4: Architecture of the smart water solution

3.2. Smart drinking water

The smart drinking water system aims at improving the detection of water leak and contamination and operating adequate measurments in case of any abnormal event.

For a rapid detection of the water leak, the water system is monitored using Automatic Meter Readings (AMRs) which measure water flow and then transmit readings to a data collector and then to a smart water platform. AMR are installed at the water supply of the central plants (Kahramaa water meter), the water tank outlet and the water supply of buidings and other facilities (Figure 5). A level water sensor is installed in the tank in order to control the tank's filling-up.

The AMR at the supply section of the central plant allows to control Kahramaa billings.
Leak in the water tank could be controlled from the AMR of Kahramaa supply, AMR at the tank outlet and the water level sensor.
Leak in the water network could be determined using the DMA method (District Metering Areas), which consists in checking the balance between the water supply (tank outlet AMR) and the totality of the water consumption in buildings (sum of AMR of buildings in the DMA).
Leak in buildings could be determined using the Artificial Intelligence (AI). For each building, the consumption profile is established from historical data. Real time readings from AMR are then compared to consumption profile. In case of a gap with between the readings and the consumption profile, a leak event is created with a severity degree related to the amount of the gap between the profile and the readings.

Water quality is controlled using S::CAN station at the outlet of the water tank. This station records the following water quality parameters: UV, Turbidity, TOC, DOC, Temperature, Color, pH, Conductivity, Temperature and Chlorine. Data are recorded at 2 minutes interval and transmitted to the Education City Platform. Data analysis is based on the deviation of readings from parameters' base-line. The deviation from the base-line indicates events related to water quality change. For the determination of the contamination origin, Saab et al. (2018) proposed to use a combination of data recorded by the Chlorine and Turbidity sensors.

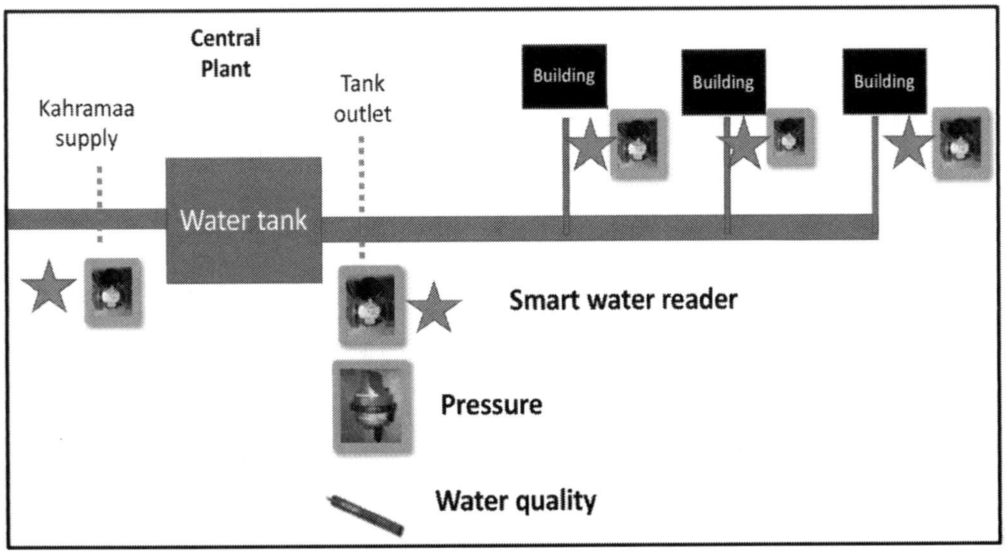

Figure 5: Smart monitoring of the drinking water system (Education City, Doha)

3.3. Irrigation system

The smart irrigation system aims at detecting water leak and optimization of the irrigation process. The leak detection system is similar to that used in the drinking water system (Figure 6).

The optimization of the irrigation service is based on the use of a monitoring system, which controls the soil moisture in the plants' root zone in order to ensure water plants' need. Sensors are used to measure the soil moisture and to transmit data to the smart water platform. The water platform conducts data analysis using historical and real-time data, weather prevision and the plants irrigation needs and then it controls the irrigation operation using electronic valves.

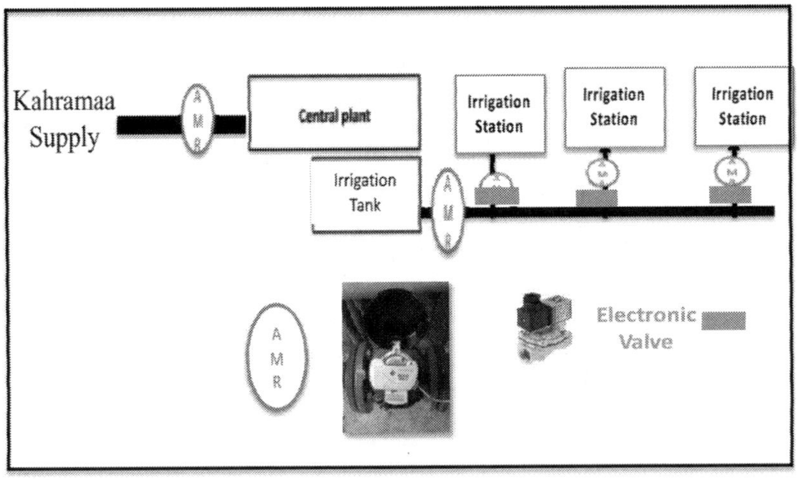

Figure 6: Smart monitoring of the irrigation system (Education City, Doha)

3.4. Fire water system

The smart fire protection system aims at the detection of water leak and the control of the hydrants pressure. The leak detection system is similar to that presented in the drinking water system (Figure 7).
Pressure sensors are installed in the hydrants for a real-time control the water pressure. In the case of a drop in the water pressure, alerts are sent to ithe management and mainteance teams.

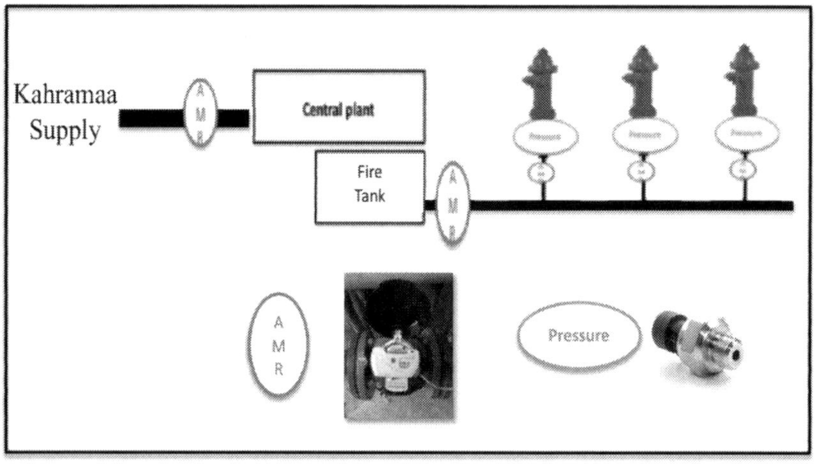

Figure 7: Smart monitoring of the fire-protection system (Education City, Doha)

4. Conclusion

This paper presented the transformation of the water system of the Education City in Doha into a smart water system, that ensures real-time monitoring of the drinking water, irrigation and fire protection systems. A smart water platform is used for data collection, data storage, data analysis and water system control.

The smart water system allows a rapid detection of the water leak. In addition, it allows a real-time control of the water quality in the drinking water system and the pressure drop in the fire protection system as well as the optimization of the irrigation process.

This system offers the possibility to inform users about the water service in the Education City and about the impact of strategies implemented for the sustainable water management. Consequently, it contributes to reinforce users' involvement in the implementation of sustainable water management strategies.

5. References

1. Abbas, O., Y. A. Rjeily, M. Sadek & I. Shahrour (2017) A large-scale experimentation of the smart sewage system. Water and Environment Journal, 31, 515-521.
2. Abou Rjeily, Y., O. Abbas, M. Sadek, I. Shahrour & F. H. Chehade (2017) Flood forecasting within urban drainage systems using NARX neural network. Water Science and Technology, 76, 2401-2412.
3. Farah E., & Shahrour I. (2017) Leakage Detection Using Smart Water System: Combination of Water Balance and Automated Minimum Night Flow. Water Resources Management, 31(15):4821-4833.
4. Lambrou T. P., Anastasiou C. C., Panayiotou C. G. & Polycarpou,M. M. (2014) A low-cost sensor network for real-time monitoring and contamination detection in drinking water distribution systems. IEEE sensors journal, 14(8): 2765-2772.

5. Liu P., Wang J., Sangaiah AK., Xie Y. 5 and Yin X (2019). Analysis and Prediction of Water Quality Using LSTM Deep Neural Networks in IoT Environment. Sustainability 2019, 11, 2058.
6. Mishra, D.R., D'Sa E.J., Mishra S. Preface (2018) Remote sensing of water resources. Remote Sens. 2018, 10, 115.
7. Saab C., Farah E., Shahrour I., Hage-Chehade F., & Ounaies, S. (2018) Use of the smart technology for water quality control: feedback from large-scale experimentation. Analog Integrated Circuits and Signal Processing, Springer, 96(2):327-335.
8. SW4EU (2017) SmartWater4Europe: Demonstration of integrated smart water supply solutions at 4 sites across Europe. European project final report. Grant Agreement no 619024. https://cordis.europa.eu/docs/results/619/619024/final1-final-report-sw4eu-v2-0-20180228.pdf
9. Shahrour I., Abbas O., Abdallah A., Abou Rjeily Y. Afaneh, A., Aljer A., (2017). Lessons from a Large-Scale Demonstrator of the Smart and Sustainable City. In A. Brdulak, & H. Brdulak (Eds.), Happy City - How to Plan and Create the Best Livable Area for the People (pp. 193-206). Springer International Publishing.
10. Storey M.V.; van der Gaag B.; Burns, B.P. (2011) Advances in on-line drinking water quality monitoring and early warning systems. Water Res. 2011, 45, 741–747.

Prediction of water consumption using Artificial Neural Networks modelling (ANN)

Elias Farah[1,2,3,*], *Amani* Abdallah[1,4], and *Isam* Shahrour[1]

[1]Laboratoire de Génie Civil et géo-Environnement, Université de Lille, 5900 Lille, France

[2]Civil Engineering Department, Lebanese University, Faculty of Engineering, Branch 1, North Campus, Ras Maska, Koura, Lebanon

[3]Civil Engineering Department, Holy Spirit University of Kaslik USEK, Faculty of Engineering, Jounieh, Lebanon

[4]Faculty of Technology, Lebanese University, Saida, Lebanon

Abstract. This paper presents an application of Artificial Neural Network models (ANN) to predict the water consumption at two scales: i) District Metered Area (DMA) located in the Scientific Campus of Lille University and ii) End user representing a restaurant inside this DMA. Data are collected from Automated Meter Readers (AMRs) that measure in near real-time the water consumption. The models are trained at both daily and hourly time intervals using historical values and the variation between the hour and the type of days. The paper shows that the ANN-based models can well predict the water consumption including peak values.

1 Introduction

Water demand prediction is widely required by water utilities for operational reasons in distribution networks and for updating the pricing policies. Nowadays, water management tools become more relevant due to the development of smart metering. The Artificial Neural Network (ANN) is one of the modelling approaches used in the hydroinformatics domain, especially for predicting water consumption [1, 2]. The ANNs are efficient in modelling complex linear and non-linear relationships without the need of implicit assumptions [3].

Many studies have been conducted to predict water demand using ANN models. [4] developed a model to forecast, at a daily basis, the water demand of a city considering meteorological indicators. It has been shown that time series data are necessary to be combined with other ANN inputs to get a reliable prediction.

Another forecasting model was presented by [5]. The study was based on statistical results and artificial neural networks to simulate the domestic water consumption for nine users. However, the developed models faced some difficulties to accurately predict the peak values. Limited research has been devoted to improve the forecasting of peak demand. [6] found that this peak is affected by the socio-economic and behavior changes of the residential sector.

This paper presents a methodology to predict water consumption at different scales using the ANN approach. Data were collected from a large scale demonstrator of the smart water system SunRise. The latter concerns the water distribution network consisting of 15 km of cast iron pipes supplying the Scientific Campus of Lille University located in the North of France which stands for a town of 25000 users. This water network is monitored by Automated Meter Readers (AMRs) that measure hourly the water consumption hourly and transmit the recorded data at daily time interval [7].

This paper presents first the methodology followed in this research including data collection and ANN modelling then discusses results obtained at daily and hourly time scales.

2 Methodology

2.1 Data collection and analysis

Water consumption data were obtained at hourly time interval from AMRs in the period from the 1st of January 2015 till the end of April 2016. This study is based on data related to (i) a DMA that comprises university faculties, administrative buildings and students' residences and (ii) a restaurant inside this DMA. Statistical analysis is conducted first in order to predict the water usage profile during the year and identify the boundaries of the study domain. Figure 1 shows the water consumption of the DMA and the restaurant at monthly time interval. It could be observed that the water consumption is

* Corresponding author: elias_farah@live.com

related to the university calendar, with a significant drop during summer holidays. This is expected since teaching and administrative buildings are closed during this period and the resident students leave the campus for summer break. In addition, the water consumption in this study is not related to the seasons or the weather variation. This result could be explained by the localization of the demo site in North of France, which has a marine west coast climate according to the Köppen Climate Classification [8] and experiences rainfall more than 150 days per year.

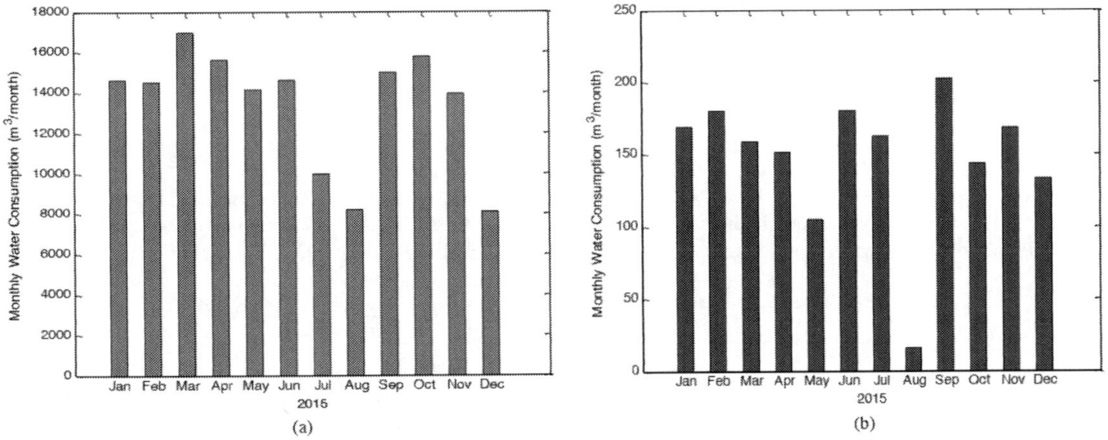

Fig. 1. Monthly water consumption in 2015 for: a) DMA and b) restaurant

2.2 Modelling approach

A feed-forward back-propagation artificial neural network is used in the numerical modelling using a Matlab script. The time series of historical water consumptions are used in the model to forecast future demand. Two main data analyses were carried out:

- The first analysis aimed at predicting daily water consumption and forecast the demand for the next day taking into account a lag of 1 day in order to study the effect of the water consumption of the day before on the actual demand.
- The second analysis was conducted to predict the water consumption for the next day with a step of 1 hour using the historical values of the day before, with a lag length of 24 hours. This lag interval is chosen due to the transmission of recorded data at 24 hours` time interval.

The input vector for the daily basis includes the day of the week DW_i (i varies from 1 to 7, i.e. Monday, Tuesday, Wednesday, etc.), the holidays HO (i.e. automn, winter, spring and summer holidays) and the special days SD (i.e. New Year's day, Easter Monday, Labor day, etc.). The input values are either 0 or 1. In other terms, for the DW_i input, if the day is a Monday, the value will be set to 1 and the others days will be assigned as 0. For the daily prediction, a new vector is added to the input consisting of the water consumption of the day before Q_{d-1}. The target data set correspond to the daily water consumption.

Analysis at hourly time interval required the consideration of the additional following parameters: a vector representing the hour of the day HD_i (i varies from 0 to 23) and the time series of historical water consumption with a lag of 24 hours Q_{h-24}.

A pre-processing step is necessary to filter the values from any abnormal event due to a reported leak or to any faults in the signal. The anomalies in time series are identified using Chebyshev's inequality [9]. This theorem guarantees that in any probability distribution, the majority of the values are close to the mean value. Using this inequality, the values are filtered to eliminate the outliers that exceed the average plus three standard deviations (i.e. 89% of the values lies within two standard deviations).

The input dataset is then divided into three sets: 70% for training, 15% for validation and the remaining 15% corresponds to the testing set. The input and target variables are normalized in the range of 0 and 1 according the following equation:

$$\overline{X} = \frac{X - X_{\min}}{X_{\max} - X_{\min}} \quad (1)$$

where X^- is the normalized input or target variables X, X_{min} and X_{max} represent the minimum and the maximum of observation values respectively. The aim of normalization is to avoid any prioritization of the variables and to remove any arbitrary effect of similarity between the objects [10].

The performances of the ANN prediction models are evaluated by the root mean square error (RMSE) and the coefficient of determination (R^2):

$$RMSE = \sqrt{\frac{1}{N}\sum_{i=1}^{N}(y_i - \hat{y}_i)^2} \quad (2)$$

$$R^2 = 1 - \frac{\sum_{i=1}^{N} (y_i - \hat{y}_i)^2}{\sum_{i=1}^{N} (y_i - \bar{y})^2} \tag{3}$$

3 Results and Discussions

3.1 Daily prediction analysis

Collected data for the DMA and the restaurant are trained using the water consumption of the previous day (i.e. lag equal to 1 day). Table 1, figures 2 and 3 show the quality of the ANN predictions.

For the DMA, a good correlation is observed between recorded and predicted values (R^2 is equal to 0.827 and RMSE is equivalent to 0.110); the ANN model predicts well the water consumption, including the peak and minimum values. The ANN model performances for the restaurant are yet better with a factor R^2 equal to 0.902 and RSME equivalent to 0.1.

Table 1. Statistical accuracy measures of ANN models at training and testing sets (daily analysis)

		Lag = 1 day	
AMR	Set	RMSE	R^2
DMA	Training	0.112	0.814
(Bulk meter)	Testing	0.110	0.827
Restaurant	Training	0.089	0.918
(End user meter)	Testing	0.100	0.902

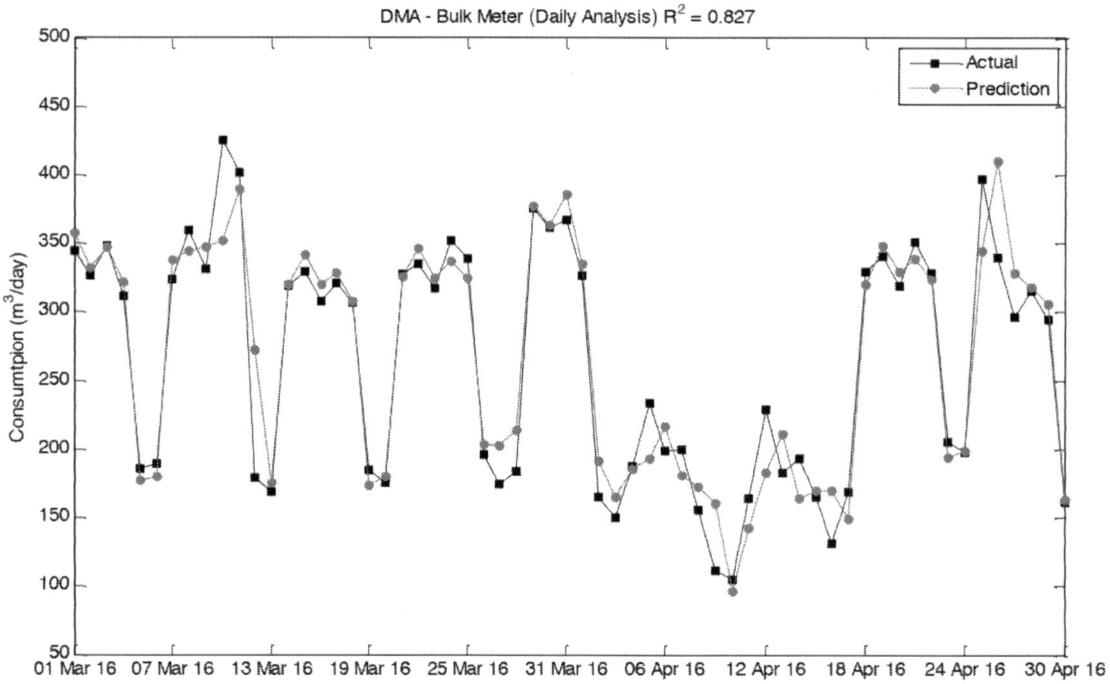

Fig. 2. Comparison between recorded and ANN prediction for the DMA (daily-time interval)

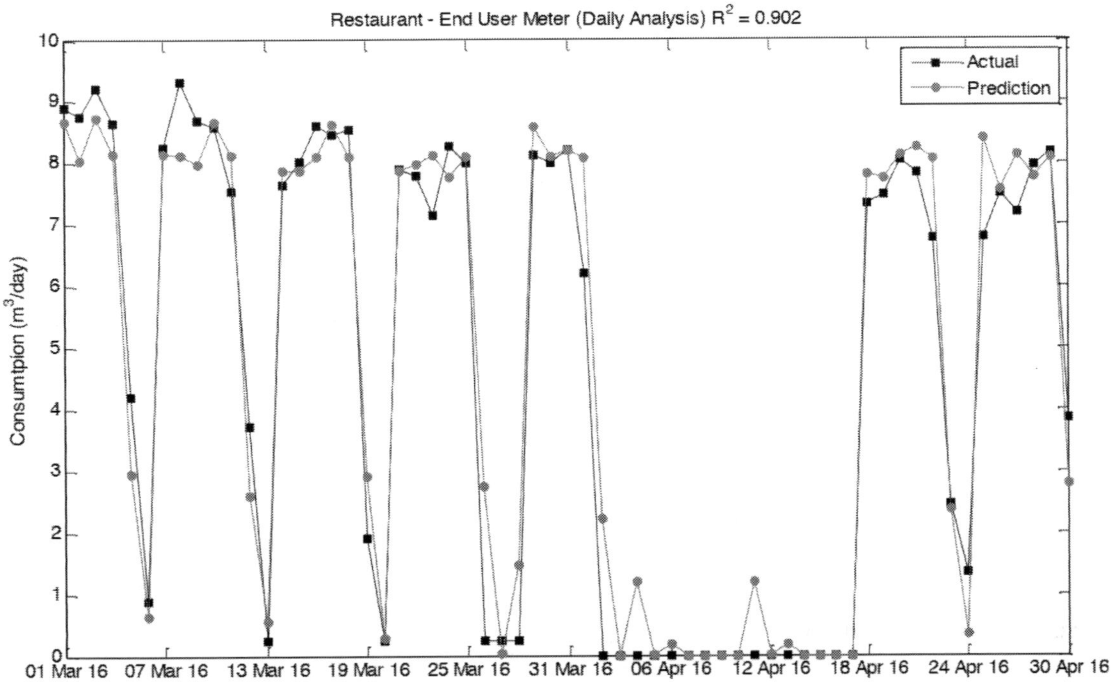

Fig. 3. Comparison between recorded and ANN prediction for the restaurant (daily-time interval)

3.2 Hourly prediction analysis

In order to model the water demand in quasi real time even, different ANN models were trained taking into account the water consumption at the hourly scale. A set of vector representing the hour of the day HD_i (i varies from 0 to 23) is added to the input matrix in addition to the day of the week, holidays and special days variables and the time series of historical values with a lag of 24 hours.

Table 2 and figure 4 show the quality of the ANN predictions. For the DMA, the ANN model predicts well the water consumption including the peak and minimum values with $R^2 = 0.825$ for the ANN testing phase. Results obtained for the restaurant are also very good with $R^2 = 0.884$ which is slightly higher than the correlation from the data of the DMA. This can be explained by the fact that data from the restaurant's water meter represent a unique profile whereas the DMA includes different behaviors of water consumption.

Table 2. Statistical accuracy measures of ANN models at training and testing sets (hourly analysis)

		Lag = 24 hours	
AMR	Set	RMSE	R^2
DMA	Training	0.062	0.857
(Bulk meter)	Testing	0.070	0.825
Restaurant	Training	0.054	0.900
(End user meter)	Testing	0.059	0.884

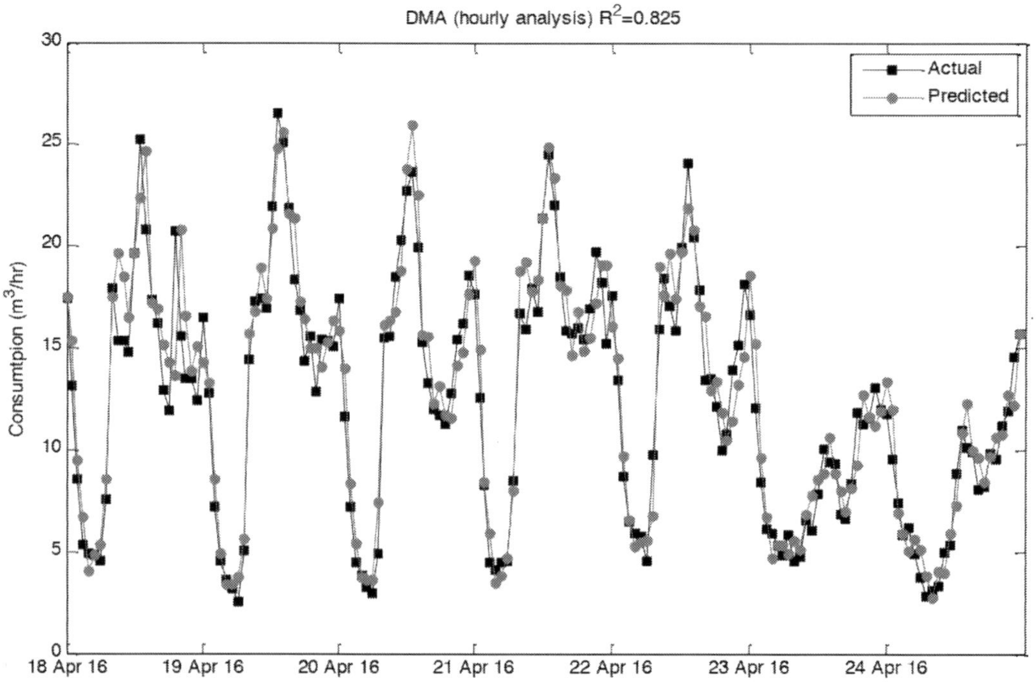

Fig. 4. Comparison between actual and ANN predicted hourly consumption values for the DMA

4 Conclusion

This paper presented the use of feed-forward back propagation ANN models for the prediction of the water consumption in the campus of Lille University at both daily and hourly time intervals. Analyses showed good performances of the ANN model for both time-scales, including peak values. For the hourly prediction, considering the historical values with a lag of 24 hours, the time and day of the week, holidays and special days, the root mean square error equals 0.07 at the DMA scale and 0.059 at the end-user one.

The ANN-based model could be easily implemented and used for short-time consumption prediction. Coupled to real-time consumption recording, they could indicate anomalies in water consumption, in particular those related to water leak. This method could also be used to estimate "recover" missing data.

References

1. Guo, G., and Liu, S., Short-term water demand forecast based on deep neural network. *1st International WDSA / CCWI 2018 Joint Conference, Kingston, Ontario, Canada* (2018)
2. Romano, M., and Kapelan, Z., Adaptive water demand forecasting for near real-ti-me management of smart water distribution systems, *ENVIRON MODELL SOFTW*, **60**, 265-276 (2014)
3. Somers, M. J., and Casal, J. C., Using artificial neural networks to model nonlinearity: The case of the job satisfaction-job performance relationship, *Organ. Res. Methods* , **12 (3)**, 403-417 (2009)
4. Al-Zahrani, M. A., and Abo-Monasar, A., Urban residential water demand prediction based on artificial neural networks and time series models, *WATER RESOUR MANAG*, **29 (10)**, 3651-3662 (2015)
5. Walker, D., Creaco, E., Vamvakeridou-Lyroudia, L., Farmani, R., Kapelan, Z., and Savić, D., Forecasting domestic water consumption from smart meter readings using statistical methods and artificial neural networks, *Procedia Eng.*, **119**, 1419-1428 (2015)
6. Day, D., and Howe, C., Forecasting peak demand-what do we need to know? *WATER SCI TECH-W SUP*, **3(3)**, 177-184 (2003)
7. Farah, E., and Shahrour, I., Smart water technology for leakage detection: feedback of large-scale experimentation, *ANALOG INTERGR CIRC S*, **96 (2)**, 235-242 (2018)
8. Peel, M. C., Finlayson, B. L., and McMahon, T. A., Updated world map of the Köppen-Geiger climate classification, *HYDROLOGY AND EARTH SYSTEM SCIENCES*, **4 (2)**, 439-473 (2007)
9. Kvanli, A., Pavur, R., and Keeling, K., Concise managerial statistics. Mason, OH: South-Western Thomson Learning (2006)
10. Riad, S., Mania, J., Bouchaou, L., and Najjar, Y., Rainfall-runoff model usingan artificial neural network approach, *MATH COMPUT MODEL*, **40 (7)**, 839-846 (2004)

ARCHITA: an innovative multidimensional mobile mapping system for tunnels and infrastructures

Federico Foria[1*], *Gabriele* Avancini[1] , *Ricardo* Ferraro[1] , *Gabriele* Miceli[1] , and *Esther* Peticchia[1]

[1]ETS Srl, Via Appia Nuova 59, Roma, Italia

Abstract. This paper presents ARCHITA: an innovative multidimensional mobile mapping system for the survey and the diagnostic of tunnels and infrastructures. The system gives the outputs of geometry and structures conditions in an integrated environment. The paper provides a brief description of the system and the successful application to support the decision-makers and the designers. Moreover, the outputs of the system are presented throughout the paper for different tunnels.

1 The heritage infrastructure in Italy and Europe

The European Union brought up the need of the renovation of the "European Core Network" (e.g. TEN-T, Trans European Network-Transport) railway to the European standards. The perspective is to ensure links between the different transport modes, eliminating existing choke point. The completion of the "European Core Network" for 2030 is an important effort for Italy, especially regards to the efficiency of rail and road connections and the completion of the "last mile" connections to ports and airports.

RFI (Rete Ferroviaria Italiana - Italian Railway Network) has been undertaking for several years an intense activity of maintenance and renewal of the infrastructure heritage. On the other hand, ETS is an engineering company with a strong background started in the railway industry.

RFI has foreseen a difficult, long and challenging work plan for the Italian infrastructure. In order to guarantee efficient investments, the mission of ETS has been to support the renewal of the strategic railway lines. In this context, ARCHITA was born. It was designed to satisfy the customer needs to reduce the time of knowledge of the infrastructural heritage and for its decision-making, focusing only on the most critical and strategic railway lines (on a small scale). ARCHITA has given fast and reliable outputs to support design solutions and works strategy (large scale) with preliminary economic evaluation through infrastructure and structures detection, maintenance estimation, slope stability assessment and clearance analyses.

* Corresponding author: federico.foria@etsingegneria.it

2 ARCHITA

2.1 Devices and configurations

ARCHITA is a multi-dimensional mobile mapping system developed by ETS consisting of linked and integrated equipment:
- Survey sensors: laser scanner, high-resolution photo, thermal image cameras, georadars;
- Positioning sensors: GPS, IMU, Odometer (they are integrated and work with a "tightly coupled" Kalman filter that accepts additional information as Ground Control Points (GCP) in post-processing. In this way, the trajectories are corrected through adjustment procedures.

ARCHITA uses the GNSS+IMU sensors to calculate the moving trajectory in a predefined reference system. The position and attitude of a particular point of the system, through the knowledge of appropriate lever-arms, is recalculated with all the related acquisitions (i.e. photographic images, thermal images and cloud points).

The precision of the system depends on the positioning precision of the GNSS system. This one depends on various factors, such as: the number, type, visibility and layout of the satellites, and the eventual absence of GPS signal (outage). In ideal conditions (open sky) and without interruption of the signal, the positioning precision of the trajectory can be equal to 2 cm.

ARCHITA carries out the survey activity with an average speed of 15-30 km/h, minimizing the impact on existing lines with shortstop of the traffic and without disconnection of the electrical tension (for railway). The system can work in two different configurations.

RAIL configuration is equipped with:
- **Laser Scanner (Leica Pegasus:Two)**, in configuration with n. 8 digital cameras, 1 IMU inertial platform, 2 profiler Z + F 9012, 2 GPS antennas, 1 optical odometer and 4 thermal imaging cameras. The two profilers, arranged at 30° and 60° to the axis of the binary track, allow a reduction of the "shadows".
- **Ground Penetrating Radar (SRS SafeRail System by I.D.S. GeoRadar)**, in the configuration of n.3 radar antennas characterized by 400 MHz central frequency.
- **Ground Penetrating Radar (by I.D.S. GeoRadar)**, in the configuration of n.1 radar antenna characterized by 600 MHz central frequency.
- **HD images (by ADTS)**, in configuration of 3 high-resolution linear cameras, 1 profiler Z + F 9012 and lighting system composed of n.16 LED light, on steel structure and aligned with the cameras.

ROAD configuration is equipped with:
- **Laser Scanner (Leica Pegasus:Two)**, in configuration with n. 8 digital cameras, 1 IMU inertial platform, 1 profiler Z + F 9012, 1 GPS antennas, 1 odometer.
- **Ground Penetrating Radar (Hi Pave System by I.D.S. GeoRadar)**, in the configuration of n.1 radar antenna characterized by 2 GHz central frequency and n.1 antenna characterized by 400/900 MHz central frequency.
- **Ground Penetrating Radar (by I.D.S. GeoRadar)**, in the configuration of n.1 radar antenna characterized by 600 MHz central frequency.
- **HD images (by ADTS)**, in configuration of 3 high-resolution linear cameras, 1 profiler Z + F 9012 and lighting system composed of n.16 LED light, on steel structure and aligned with the cameras.

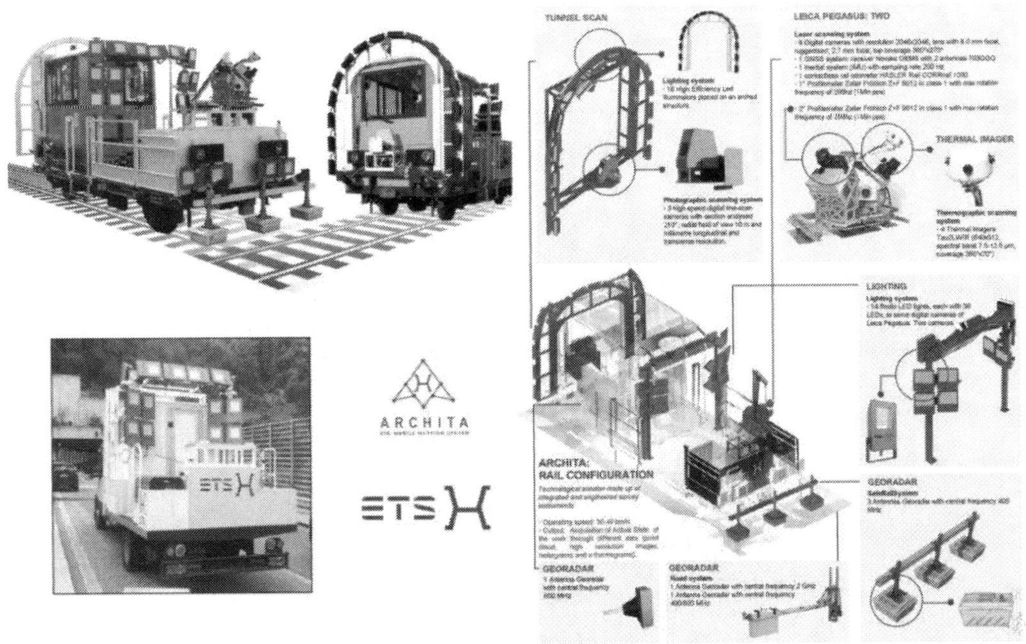

Fig. 1. ARCHITA equipment (RAIL/ROAD configuration).

ARCHITA can obtain a series of integrated and connected information in a single passage, such as:

- *Geometry* (Figure 2): georeferenced 3D point cloud. In post-processing, it's possible to obtain the track geometry and the correspondent CAD models (i.e. tunnel shape, clearance analysis, track, electric supports and contact line).

Fig. 2. Galleria Sipicciano and Galleria Olmata (Italy): outputs from 3D point cloud.

- *Ballast* (RAIL configuration, Figure 3): 400-600 MHz georadar, for the definition of the thickness, material alterations, identification of cavities, and presence of water.
- *Lining* (Figure 4): 600 MHz georadar, for the definition of the thickness, identification of cavities, material alterations and presence of water.

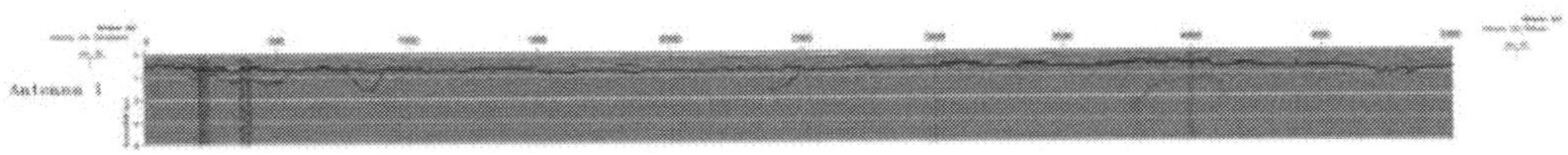

Fig. 3. Roma-Cassino railway line (Italy): GPR survey on the ballast.

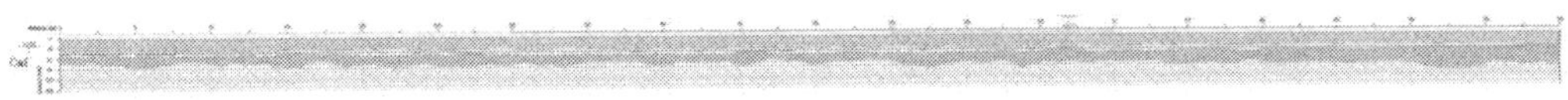

Fig. 4. Galleria Sipicciano (Italy): GPR processing survey on the tunnel lining.

- *Lining condition*: high-definition photo (Figure 5) and thermographic analysis (Figure 6) synchronized with the 3D point cloud.
- *Pavement* (ROAD configuration): 2.0 GHz georadar, suitable frequency for the definition of the asphalt thickness and 400/900 MHz georadar, suitable frequency for the definition of the subgrade thickness.

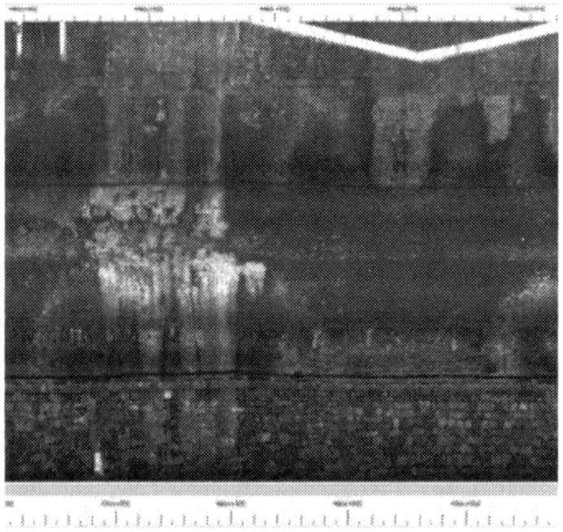

Fig. 5. Galleria Montorso (Italy): HD photographich images.

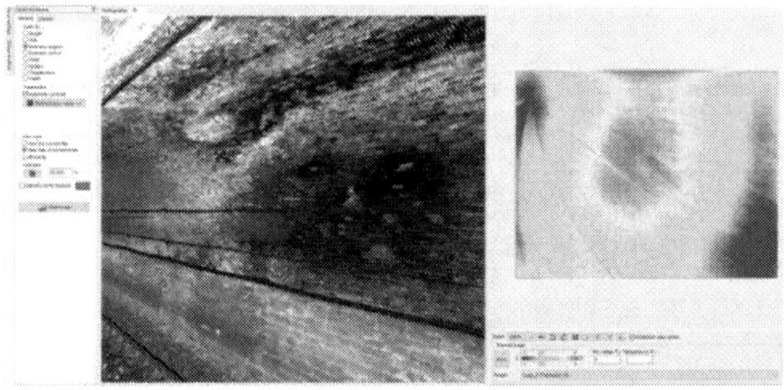

Fig. 6. Galleria Montorso (Italy): 3D point cloud synchronized with thermographic images.

2.2 Product and software development

ARCHITA has passed through a severe process of development (still on-going). Leica Geosystems customized the laser scanner adding n.1 profiler, n.1 GPS antenna and n.4 thermal cameras. ADTS customized the Tunnel Scan adding n.1 profiler and n.1 GPS antenna.

Bruno Kessler Foundation (FBK) coordinated the integration between the Laser Scanner (Leica Geosystems) and the Tunnel Scan (ADTS). Bruno Kessler Foundation (FBK) has demonstrated successfully the integration through a field campaign on the railway line Attigliano-Viterbo (the integration tests were performed in the Sipicciano I and Sipicciano II tunnels in Italy). This study proved that the 3D point clouds of the two instruments have a maximum deviation of 1.5 cm.

The development has also covered the elaboration software: Leica Geosystems customized the post-processing software based on the new configuration of the product. ADTS has been customizing the post-processing software based on the specific requests of ETS and the clients.

The style for the names is First Names, typed in italic 10-point Times, then Last Name, typed in 10-point Times, with a comma after all except the last author, which is separated by comma + “and”. Do not use academic titles.

Fig. 7. Galleria Sipicciano (Italy): results of point cloud integration (Tunnel Scan - Pegasus:Two).

3 Application: geometric survey and clearance analysis for DTP Roma, Italy

3.1 Problem and requirements

ETS has been in charge to investigate for the DTP Roma (RFI Lazio railway district), which railway line between Rome and Naples is more convenient to renew. This feasibility study and the related project are part of the Scandinavian-Mediterranean Corridor, which represents the north-south axis of the Trans-European Transport Networks (TEN-T).

The project required: the geometric survey of the railway infrastructure, the widening of the clearance to ensure the strengthening of the railway, the resolution of the leakages, the mapping and backfilling of the cavities along the line and the execution of the works with respect to a working line. ETS has carried out a geometric and knowledge detection of about 1200 km of the railway line (built in the early 1900s) of which 170 km of tunnels (mainly with masonry lining), overhead lines and railway stations.

3.2 Survey of geometries and lining conditions

ARCHITA has been used to determine the inner geometry (Figure 8) in order to support the check of the clearance conditions to design the milling of the existing lining and, consequently, the appropriate reinforcement works.

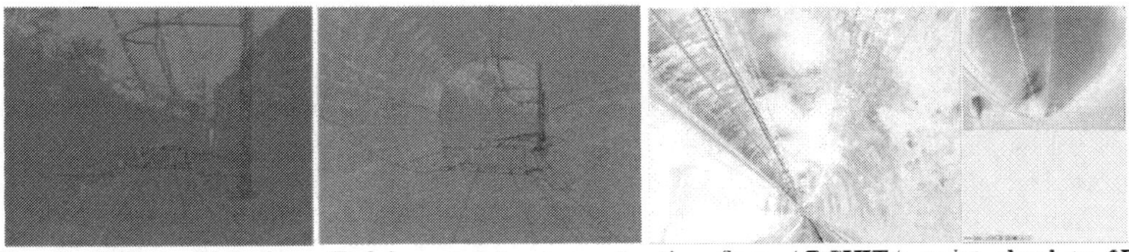

Fig. 8. Galleria Sipicciano (Italy): geometry reconstruction from ARCHITA point cloud to 3D digitalization and thermographic result.

In Figure 9, it's depicted the output of high-resolution photo to check the condition and the humidity content of the lining. In the same figure, it's shown a georadar elaboration to determine the thickness of the lining. The thicknesses have been verified and calibrated based on traditional destructive tests.

The ARCHITA outputs have been integrated in a BIM approach through all the stages of survey and design (Figure 10). ARCHITA has been used to check both ballast and lining thicknesses and condition

All these outputs support the decision-maker towards the more convenient line to renew and the designers towards the more appropriate interventions.

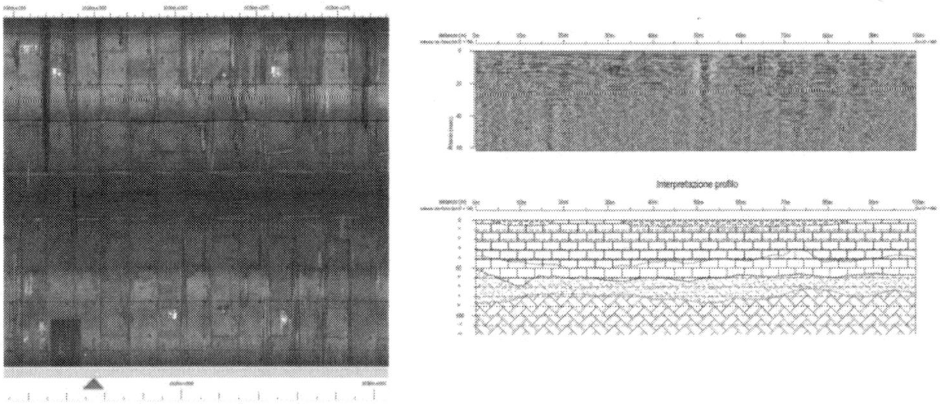

Fig. 9. Galleria Costa dei Rosi (Italy): high-resolution photo for the defects detection (left) and GPR for lining thickness (right).

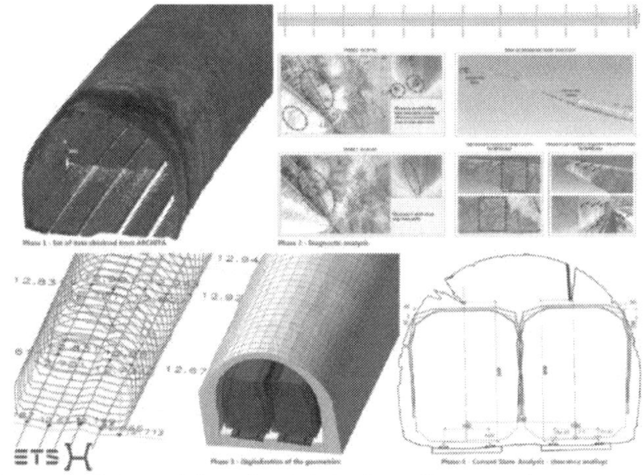

Fig. 10. Galleria Olmata (Italy): workflow for "BIM oriented" approach.

4 Conclusions

The underground space has already been widely used in the past centuries. The number of existing tunnels requiring renewal, monitoring or maintenance, increases day by day. For this reason, ETS has been developing ARCHITA. The system guarantees a quick and reliable integrated survey of the existing infrastructure to help design and feasibility studies.

Actually, ARCHITA has been deployed in more than 400 km of tunnels and 2000 km of infrastructure in Italy. One case has been briefly presented. Nevertheless, ARCHITA is proving that still a lot needs to be done to achieve a clear, objective and repeatable way to survey and manage the existing tunnels and infrastructure.

References

1. M. Gavilán, F. Sánchez, J.A. Ramosand, O. Marcos, The 6th International Conference on Structural Health Monitoring of Intelligent Infrastructure (2013)
2. G. Loprencipe, L. Moretti, T. Pestillo, R. Ferraro, Sustainability, E **10**, 3145 (2018)
3. R. Montero, J.G. Victores, S. Martínez, A. Jardón, C. Balaguer, Automation in Construction, E **59**, 99–112 (2015)

Progresses on the intelligent construction and operation maintenance of urban under-lake tunnels

Jingfeng Wang[1,2], *Guang* Liu [1,2,3*], *Zhaodong* Ding[1,2], *Hanbing* Bian[4], *Huayan* Yao[1,2]

[1] School of Civil Engineering, Hefei University of Technology, Hefei, China

[2] Anhui Key Laboratory of Civil Engineering Structures and Materials, Hefei University of Technology

[3] Key Laboratory of Rock Mechanics in Hydraulic Structural Engineering (Ministry of Education), Wuhan University, Wuhan, China

[4] Laboratoire de Génie Civil et géo-Environnement, Université de Lille, 5900 Lille, France

Abstract. China is currently at an important stage of urbanization. In recent years, the number and scale of under-lake tunnels in southern cities of China are growing consistently, which brings opportunities and challenges in intelligent construction and operation maintenance. Due to the long length of the under-lake tunnel, the large volume of concrete to be cast, the complex environment of the buried section and the underground environment of the lake bottom, the hydration heat and stress crack have been the main challenges of the under-lake tunnel engineering in crack control and waterproofing aspect. Simultaneously, with the development of big data, intelligent construction, information technology, and other techniques, these techniques are gradually applied to the whole life cycle of tunnel engineering including survey, design, construction, operation, and maintenance. The construction of the lake tunnel project is setting off a new round of technological innovation. This paper systematically summarizes the technical breakthroughs and application of technologies such as intelligent construction and information technology in the construction of under-lake tunnels in China, and puts forward reasonable suggestions on the key technologies of intelligent construction and operation maintenance for urban under-lake tunnel engineering.

* Corresponding author: guang_liu@hfut.edu.cn

1 Introduction

The construction of under-lake tunnels has set off a new upsurge in the Chinese urbanization process. During the past decades, a number of under-lake tunnels are constructed in Chinese metropolises, such as Nanjing, Wuhan and Suzhou. There are many well-known under-lake tunnels in China, for instance, East lake, Tai lake, Xuanwu lake, and Dushu lake tunnels. The detailed information for some of the well-known under-lake tunnels in China is listed in **Table 1**.

Table 1. Some of the well-known under-lake tunnels in China

Tunnel name	Overall length (km)	Width (m)	Construction state
East lake tunnel	10.6	60-70	Completion
Tai lake tunnel	10.8	43.6	Under construction
Xuanwu lake tunnel	2.7	32.0	Completion
Dushu lake tunnel	3.5	30.2	Completion

The under-lake tunnels have an important effect on the optimization of the urban road network system. The commuting time between two sides of the lake can be considerably reduced with the construction of under-lake tunnels. The commuting time from Liyuan to Guanggu which are two places on the two sides of the East lake shortens to 10 minutes from one hour after the East lake tunnel is built.

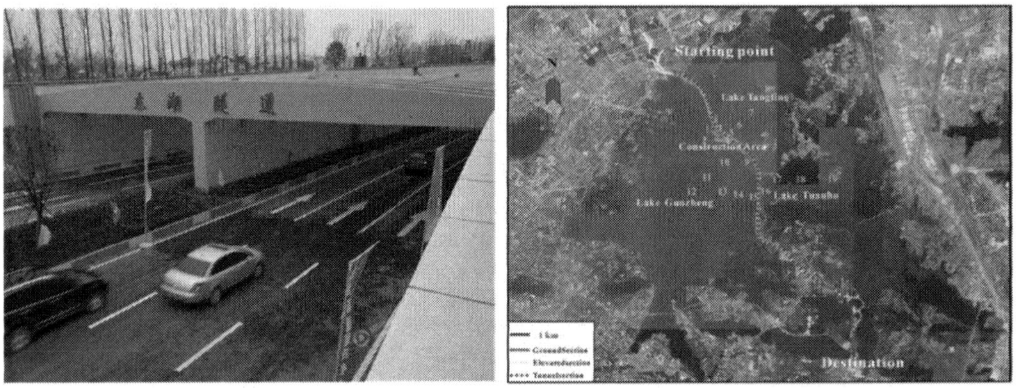

(a) (b)

Figure 1. Figures of East lake tunnel in Wuhan. (a) The East lake tunnel opens to traffic. (b) A sketch map for the route of the East lake tunnel (Wang et al., 2016).

Due to the complexity and uncertainty of the under-lake tunnels, a large amount of data analysis and prediction are needed during the construction process. The intelligent algorithms, for example, artificial neural networks (ANNs), support vector machine (SVM) and Gaussian Process (GP), can be used to select the modeling parameters and predict the mechanical response of the under-lake tunnels (Darabi et al., 2012; Mahdevari et al., 2013; Li et al., 2017). A spreadsheet optimization technique coupled with the trained neural network model is used to study the reliability assessment of the settlement serviceability limit state by Goh and Hefney (2010). Recently, Liu et al. (2019) adopted a genetic algorithm and combined covariance Gaussian process regression coupled algorithm to execute displacement back analysis and predict displacement of surrounding rock in the initial monitoring period of tunnel construction. This prediction of surrounding rock displacement also can be extended to the long-term monitoring period after the training samples are collected.

This paper first reviews the main problems faced in the current construction of the under-lake tunnels. Then the application of smart technology in the tunnel construction process is studied. A programmatic framework for the application of smart technology in the construction of the under-lake tunnels is established.

2 Major challenges for urban under-lake tunnels

There are many challenges in the whole life cycle of the construction and operation of urban under-lake tunnels. Due to limited space, this research mainly elaborates on leakage and concrete cracking.

2.1 Water leakage in the under-lake tunnels

A remarkable feature for the under-lake tunnels is the tunnels need to stretch under the lake, and the burial depth of the tunnels is highly dependent on the geological conditions of the bottom sediment in the lake. Therefore, for high permeability soils on the lake bottom, the tunnels may be subjected to great risk in water leakage.

As a matter of fact, the current status for the waterproofing of underground tunnels in China is not optimistic. This issue may be more severe for the under-lake tunnels because this underground structure has direct interaction with the lake. According to the statistics of China Building Waterproof Association (2013), the leakage rate of underground engineering in China is as high as 80%. Compared with the collapsed construction accidents during the construction process, the underground leakage is generally slow and concealed. Therefore, the underground leakage attracts less attention of all parties until a devastating accident happened.

In 2013, China Building Waterproof Association and Beijing Zero Market Research and Analysis Company (2013) jointly released a sample survey of the National Building Leakage Survey Project Report, involving 1777 underground building samples from 28 cities. In the survey, 1022 underground structures had leakage problems in varying degrees. The leakage rate reached 57.51%, in some cities such as Chongqing, Wuxi and other places, the leakage rate of underground buildings is as high as 100%. According to statistics, the annual maintenance cost of water leakage in underground space in China accounts for more than 50% of the total maintenance cost. The problem of water seepage and water leakage in urban underground structures has become a common problem in construction.

(a) (b)

Figure 2. Leakage on the sidewall of an underground passage for the motor vehicles in Hefei, Anhui. (a) An underground passage at Luyang district. (b) An underground passage at Shushan district.

2.2 Cracking of the concrete

The geological environment of the lake bottom tunnel project is significantly different from other construction projects. Due to the Huge pouring amount of concrete and complex underground environment at the bottom of the buried section, the crack propagation caused by temperature stress and uneven deformation can be found in underground structures. Cracking of tunnel concrete is a common problem in underground engineering (see Figure 3).

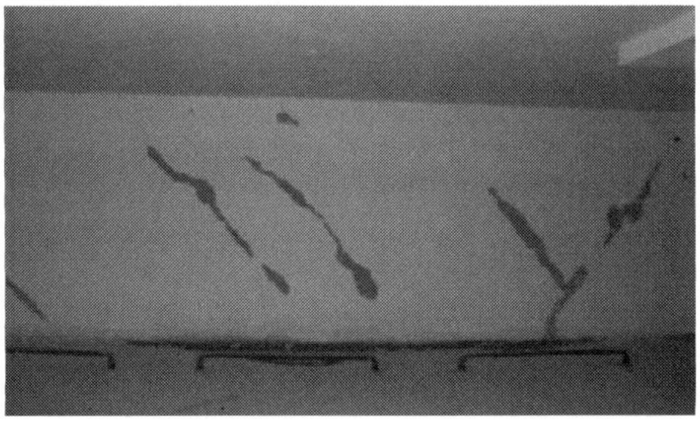

Figure 3. Cracks on a concrete shear wall

There are two main reasons which cause the cracking of underground tunnels. One is that the uneven distribution of hydration heat of mass concrete, which causes temperature cracks. The other is due to the non-uniformity properties of the foundation soil, resulting in uneven settlement or deformation at different locations for the underground structure. It can be found from Figure 4 that the damage is concentrated on the sides of the sidewall once the left side has a vertical settlement of 10 mm. This damage area is very likely to cause crack propagation in the concrete.

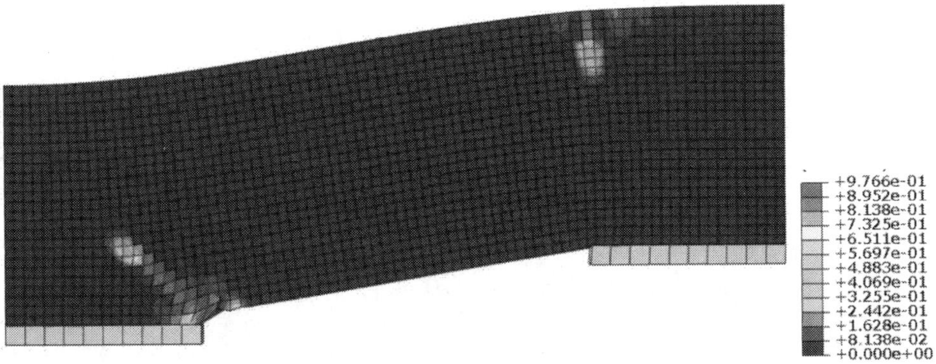

Figure 4. Damage distribution when the left the sidewall in an underground structure has a vertical settlement of 10 mm. The damage degree is represented by a dimensionless damage factor shown the legend. The damage factor ranges from 0-1. No damage is indicated by 0 and full damage indicated by 1.

In addition, concrete cracking and leakages are two coupled physical processes (Davies et al., 2014; Liu et al., 2019). It is well known that uneven external forces (water pressure, earth pressure, frost heave force, insufficient bearing capacity, earthquakes, etc.) in the underground environment are applied to the structure, causing uneven deformation or even cracking of the structure, resulting in a large localized waterproof layer. This effect is more pronounced when the tunnel encounters an inhomogeneous deformation. Once the stress exceeds the allowable value, the internal microcracks initiate, propagate, and extend to form a series of fractures, which further destroys the waterproof structural system. Meanwhile, it is worth noting that the long-term leakage also accelerates the failure of concrete due to the strength degradation of concrete under water immersion conditions.

3 Application of smart technology

At present, technologies such as big data and artificial intelligence are developing rapidly. In the construction and operation of the under-lake tunnels, these smart technologies should be used for intelligent monitoring and information management.

As mentioned above, leakage and concrete cracking are two important problems in the construction of lake tunnels. These two problems exist in the entire life cycle including construction, construction, operation, and maintenance stage of the under-lake tunnel. These problems have a long time span, and they are difficult to detect. Therefore, we can make full use of the current smart technology for automated detection and information analysis, and implement intelligent monitoring of the life cycle for the under-lake tunnels.

As shown in Figure 5, the application framework can be divided into three parts: the data acquisition layer, the analysis layer, and the control layer. The data acquisition layer mainly collects the displacement, deformation, temperature, image and other data on the key positions in the stage of construction, operation and maintenance through high-precision cameras and high-sensitivity sensors. The analysis layer uses the collected displacement, stress and other data as learning samples through the neural network, machine learning, deep learning algorithm. After the sample training, the analysis layer can predict the displacement and stress in the next stage. The analysis layer can also process image data to identify locations of cracks and leakages, which further provides guidance for manual intervention. The control layer performs manual interventions to prevent leakages and crack propagation. The chemical grouting, spraying waterproof coatings are two common interventions for the leakage and cracking problems. In addition, the control layer will optimize the construction plan and building materials based on current monitoring data.

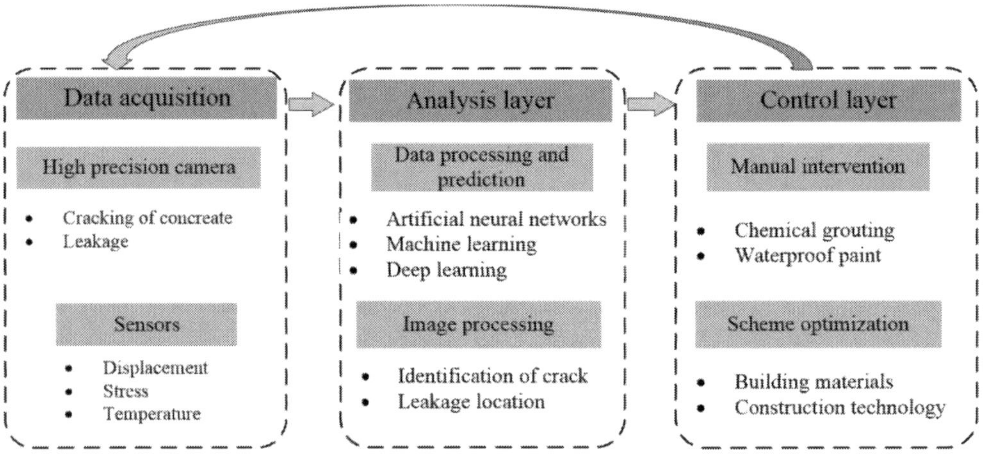

Figure 5. An application framework for smart technology in under-lake tunnels

4 Conclusions

China is at a critical stage in the process of urbanization, and the problem of urban traffic congestion is gradually emerging and becoming increasingly serious. The construction of the lake tunnel not only improves the urban road network system but also becomes a unique landscape of the city.

Considering the special geological environment of the lake tunnel, the lake tunnel often has problems such as water seepage and concrete cracking. Smart technology has been successfully applied in data processing and tunnel deformation prediction. This article proposes data acquisition through high-definition cameras, sensors, and other detection devices, and then uses intelligent algorithms for data analysis and processing. On the basis of that, the leakage and concrete cracking in the whole life cycle of tunnel construction and operation can be controlled.

Acknowledgements

This research is supported by the Fundamental Research Funds for the Central Universities of China (No. PA2019GDPK0089), the Open Fund of the Key Laboratory of Rock Mechanics in Hydraulic Structural Engineering, Ministry of Education, Wuhan University

(No. RMHSE1604), and Anhui Natural Science Youth Fund (No. 1908085QE216). These supports are gratefully acknowledged.

References

1. Davies O, Rouainia M, Glendinning S, et al. Investigation of a pore pressure driven slope failure using a coupled hydro-mechanical model[J]. Engineering geology, 2014, 178: 70-81.
2. Liu G, Sun WC, Lowinger SM, Zhang ZH, Huang M, Peng J. Coupled flow network and discrete element modeling of injection-induced crack propagation and coalescence in brittle rock. Acta Geotechnica. 2019;14(3):843-868.
3. Darabi A, Ahangari K, Noorzad A, et al. Subsidence estimation utilizing various approaches–A case study: Tehran No. 3 subway line[J]. Tunnelling and Underground Space Technology, 2012, 31: 117-127.
4. Li S, He P, Li L, et al. Gaussian process model of water inflow prediction in tunnel construction and its engineering applications[J]. Tunnelling and Underground Space Technology, 2017, 69: 155-161.
5. Mahdevari S, Haghighat H S, Torabi S R. A dynamically approach based on SVM algorithm for prediction of tunnel convergence during excavation[J]. Tunnelling and Underground Space Technology, 2013, 38: 59-68.
6. Goh, A.T.C., Hefney, A.M., 2010. Reliability assessment of EPB tunnel-related settlement. Geomech. Eng. 2 (1), 57 –69.
7. China Building Waterproof Association, Beijing Zero Market Research and Analysis Company. National Building Leakage Survey Project Report in 2013, 2013.

Study on Risk Classification of Goaf Based on RS-SVM

Haiping Yuan[1,3], *Biaohua* Le[1], *Yubo* Wang[1], *Chenghao* Chen[1], and *Hanbing* Bian[1,2]

[1]School of Civil Engineering, Hefei University of Technology, hefei anhui 230009, China
[2]Laboratoire de Génie Civil et géo-Environnement, Université de Lille, 5900 Lille, France
[3]State Engineering Laboratory of Highway Maintenance Technology, Changsha University of Science and Technology, Changsha, China

Abstract. According to the uncertainty and concealment of the risk of goaf, a risk classification model of goaf is constructed based on rough set (RS) knowledge and support vector machine (SVM) theory. In this paper, based on statistical analysis and measured data, nine parameters including mining method, empty area excavation depth, goaf height, maximum exposed area of empty area, maximum exposure height, maximum exposure span, pillar condition, empty volume and treatment rate are selected as the main influencing factors. The RS theory is used to reduce the sample, and SVM is compiled by Matlab. The one-to-one method is used to construct the binary classifier to realize the multi-class classification algorithm of goaf. Finally, a SVM model for evaluating the risk level of the goaf is obtained. The research shows that: based on RS theory, SVM has a good effect on the hazard classification of the goaf iron ore mine, and the difference with the actual situation is 13.3%. The research results have certain theoretical significance and guiding role for the safe mining of an iron mine in Eastern China.

1 Introduction

Because the mine uses the empty field method for mining, it leaves a large number of goafs. With the continuous consumption of mineral resources, the mine production activities are increasing, the mining depth is gradually deepening and the scale is gradually increasing. The goaf poses a more serious threat to the safety of the mine, and the phenomenon of rock stratum displacement and crack deformation is gradually jeopardized. The danger of the goaf has seriously affected the safety of the entire mine and has become an important factor in evaluating mine safety [1-3].

The risk assessment of goaf is still a key issue. At present, domestic and foreign scholars have done a lot of research on the risk of goaf. Chen Zhiqiang[4] established a two-level evaluation model for fuzzy comprehensive evaluation. According to the weight set and evaluation matrix, the fuzzy comprehensive evaluation of the goaf was completedaccording to the second-level evaluation model. Xu Xiaoding[5] constructed the pillar-rock beam support system in the plastic zone, and constructed the cusp catastrophe model of the support system by using energy theory, and obtained the instability judgment conditions related to the support system, so as to evaluate the stability of the goaf. Zhang Yaoping[6] used numerical simulation to simulate and predict the formation process of the empty area and the stability of the goaf. Hu Gaojian[7] established a three-dimensional discrete fracture network model of structural plane, studied the degree of cutting fracture of surrounding rock, and obtained the stability analysis result of complex goaf by using Mathews stability diagram method. Li Junjun[8] used adaptive genetic algorithm to globally optimize the initial weight and enthalpy of BP network, and predicted the risk level of goaf. Based on the theory of set pair analysis, Xiao Chao[9] constructed a pair of sampled measured index values that affect the stability of the goaf and the stability criteria of the goaf, and described the pair relationship through the relationship degree and the stability of the gab mine goaf.classification. Wang Wei[10] established a stability evaluation index system for goaf,applied matter-element analysis theory, and established a matter-element extension model for the evaluation of goaf stability through steps of matter element structure and correlation function.

Due to the complex geological environment in the goaf and the uncertainties of the stability of the goaf, many studies on the stability of the goaf have certain limitations. In this paper, the rough set (RS) is used to reduce the sample attributes, and the influencing factors are selected based on the actual situation of an iron mine in Eastern China. The support vector machine (SVM) is used to classify the risk of the goaf, which is of great significance to mine safety.

2 Basic theory of RS-SVM

RS theory proposed by Pawlak et al[11-12] is a new mathematical tool to deal with the problems of ambiguity and inaccuracy. When dealing with redundant information, the reduction process has the following definition.

SVM is a new machine learning method proposed by Vapnik et al[13]. Based on statistical learning theory. The basic theory is as follows:

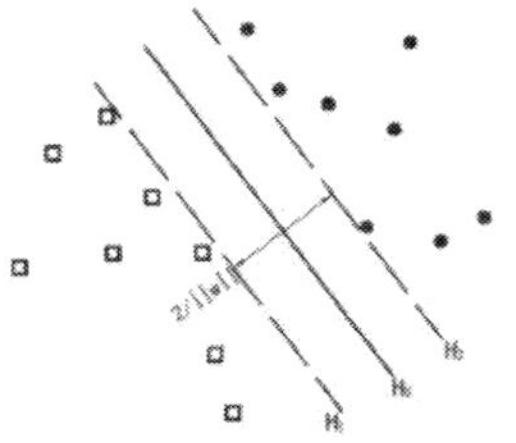

Figure 1 Optimal classification super-planar

In the case of linear separability, Suppose there is a training sample (x_i, y_i), $x \subset R^n$, $y = \{+1, -1\}$, $i = 1, 2, \ldots, l$. As shown in Figure 1, the classification interval is $\frac{2}{\|w\|}$, making the interval maximum equivalent to $\frac{1}{2}\|w\|^2$ minimum. Considering the existence of an error, using a slack variable $\xi > 0$, it is written as:

$$\min \ \frac{1}{2}\|w\|^2 + C\sum_{i=1}^{l}\xi_i \tag{1}$$

Where C is a normal number and the Lagrange function is established:

$$L = \frac{1}{2}\|w\|^2 + C\sum_{i=1}^{l}\xi_i - \sum_{i=1}^{l}\alpha_i(y_i(<w, x> + b) - 1 + \xi_i) - \sum_{i=1}^{l}\gamma_i\xi_i \tag{2}$$

In the nonlinear case, the partial derivative of the Lagrange function to the optimization target w, b, ξ is 0. According to the Kuhn-Tucker condition and the Mercer theorem, the kernel function $K(x_i, x_j)$ is selected to map the data to a high-dimensional feature space, and then in the high-dimensional Linear classification is performed in the feature space to obtain the dual form of the following quadratic programming:

$$\begin{cases} \max \ \sum_{i=1}^{l}\alpha_i - \frac{1}{2}\sum_{i=1}^{l}\sum_{j=1}^{l}\alpha_i\alpha_j y_i y_j K(x_i, x_j) \\ \sum_{i=1}^{l}\alpha_i y_i = 0 \\ 0 \le \alpha_i \le C, i = 1, 2, \ldots, l \end{cases} \tag{3}$$

Solving the optimal solution of α from the quadratic programming problem, and selecting the support vector of $\alpha \neq 0$ to obtain:

$$b = y_i - \sum_{i=1}^{l}\alpha_i^* y_i K(x_i, x_j) \tag{4}$$

Finally, the optimal classification function of SVM can be constructed:

$$f(x) = \mathrm{sgn}(\sum_{i=1}^{l}\alpha_i y_i K(x_i, x) + b) \tag{5}$$

Where, the kernel functions is follow:

$$K(x, x_i) = \exp\left(-\frac{\|x - x_i\|^2}{2\sigma^2}\right) \tag{6}$$

3 Engineering applications

3.1 Project overview

According to the current statistics, there is no less than 94 goafs in the upper plate area, F2 lower plate and east area of an iron mine in the east, including 168 sub-goafs. The total volume of the goafsis about 3.67 million m3, of which the eastern area is about 2.03 million cubic meters, accounting for 55.3% of the total volume of the empty area above 200m; most of the empty area is concentrated in the horizontal range of 200-300m, the total volume of the empty area is about 1.63 million m3, accounting for nearly 44.8% of the total volume of the empty area above 200m. The empty area is mainly formed by shallow hole mining method, and the shallow goaf accounts for nearly 75.8% of the total volume of the empty area above 200m.

3.2 Influencing factors of goaf

There are many factors affecting the stability of the empty area, such as engineering and hydrogeology, components of the ore block, depth of burial of the ore body, reasonable design of the pillars in the isolated pillar and the stope, vibration and impact of the blasting on the ore body and surrounding rock. The length of the formation of the goaf, the shape ofthe goaf, the exposed area, the exposed height, the thickness of the roof and the bottom layer, the treatment and strength of the empty area, the surrounding rock strength of the empty area,, the scale and distribution of the ground surface and joint fissures , the development of geological structures such as faults, and so on. Moreover, the influence of each factor on the stability of goaf in a specific mine is quite different[14-15]. According to the data collected and mastered by an iron mine in the east of +200m and above, the evaluation criteria of the stability of the empty area include mining method (C_1), empty area excavation depth (C_2), goaf height (C_3), mining The maximum exposed area (C_4), the maximum exposure height (C_5), the maximum exposure span (C_6), the pillar condition (C_7), the volume of the goaf (C_8), and the treatment rate (C_9). SVM can't process qualitative indicators when learning sample data, so it needs to be quantified by qualitative indicators.

3.3 Selection of samples and kernel functions

At present, there are 168 sub-empty areas in an iron ore sub-space in the east. The training samples are randomly selected by stratified sampling. As shown in Table 1, the previous 33 sequences are used as training samples, and the rest are used as prediction samples. When selecting the kernel function, considering that the influencing factors of the goaf are numerous and the dispersion is large, the linear kernel function can not effectively classify the risk of the goaf; the Sigmoid kernel function can classify well, but the parameter value requirements Higher, and the effect is not ideal under small samples. The polynomial kernel function has global characteristics, while the RBF kernel function has a wide range of applications and a wide convergence domain[16]. In this paper, polynomial kernel function and RBF kernel function are used to predict the risk.

Table 2 Samples of goaf danger classification

Serial number	Excavation depth (m)	Mining method	Gob height	Maximum exposed area (m^2)	Maximum exposure height (m)	Maximum exposure span (m)	Pillar situation	Measured volume (m^3)	Governan-ce rate	Actual level
1	166	1	12	8114.7	12	62	1	48684	1	I
2	336	1	14	2615	14	46	2	33085	1	I
3	308	1	15	2519	15	50	2	21721	1	I
4	160	2	17	3883	4.2	26	3	33805	3	II
					......					

3.4 Sample data processing and attribute reduction

The sample data size has a large impact on the SVM model, which leads to a decrease in the accuracy of the classification results. Therefore, in order to ensure the stability of the numerical operation and the convergence of the model network, and to eliminate the influence of the dimension, the sample data needs to be Normalized according to equation(7), namely:

$$S_{ij} = \frac{x_{ij} - \min(x_j)}{\max(x_j) - \min(x_j)} \quad (7)$$

Where: x_{ij} is the sample data before normalization; S_{ij} is the normalized sample data; $\min(x_j)$ is the minimum value in the measured data of the index; $\max(x_j)$ is the maximum value in the measured data of the indicator.

In the process of reduction using rough set theory, in order to narrow the information system and improve the clustering ability of the evaluation system, it is necessary to discretize the normalized data.This paper uses the equidistant division method to discretize.The spatial dimension and information loss after attribute discretization should be minimized, therefore, take three breakpoints when discrete, and establish an information decision table.

Based on the information decision table, the redundant information in the data is removed, and the data dimension is reduced. The information decision table shows that the training samples 2 and 3,16 and 18, 20 and 24, 23 and 28 overlap, thus deleting the duplicate Training samples 3,18, 24, 28. At the same time, the attribute reduction is performed by the deletion method, and the reduction set is $\{C_2, C_5, C_8\}$.

3.5 Establishment of SVM multi-class classification model

At present, the common methods for multi-class classification by SVM include one-against-One, one-against- Rest, binary tree method and directed acyclic decision graph method[17], which are all based on SVM binary classification. In this paper, the SVM multi-class classifier is constructed by one-against-One. This method constructs $k(k-1)/2$ sub-classifiers in k-class training samples. Each sub-classifier only has two kinds of training samples in k class. Train between them, then use the voting method to examine the test samples one by one, and select the category with the most votes as the final output category of the classifier.

The RS-SVM multi-class classification model program is written based on matlab. The parameters are determined by the trial algorithm as follows: the kernel parameter of the polynomial kernel function $\gamma = 0.5$ and $q=2$, the RBF kernel function $\sigma^2=0.3$, and the penalty parameter $C=2.5$. The predicted results are shown in Figure 2.

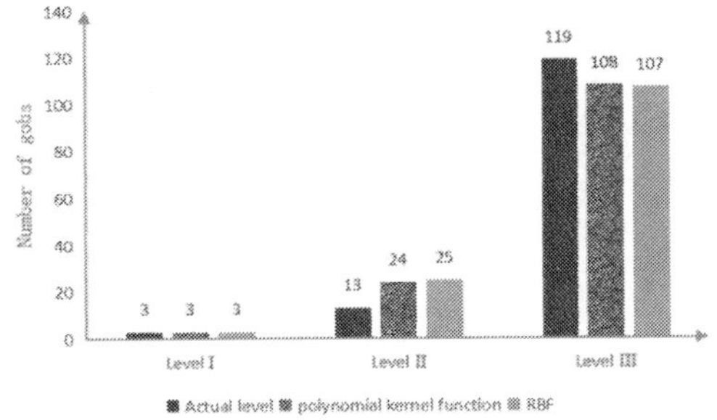

Figure 2 Comparison of risk classification of goaf

When using SVM for grading, the influence of the kernel function parameters and the penalty parameters is very obvious. The error of the final evaluation can be obtained by debugging the parameters, as shown in Figure 3 and Figure 4.

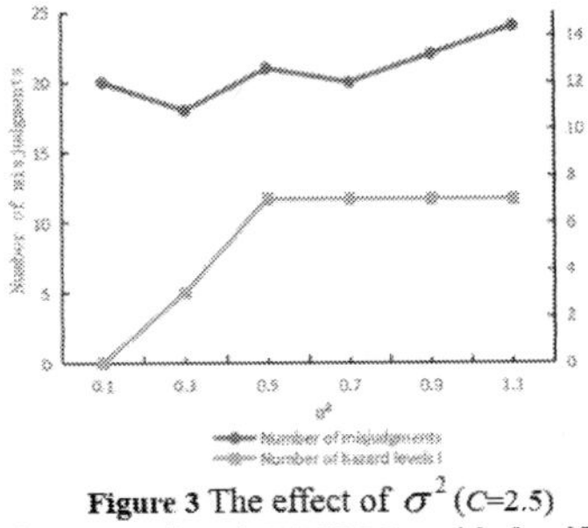

Figure 3 The effect of σ^2 (C=2.5)

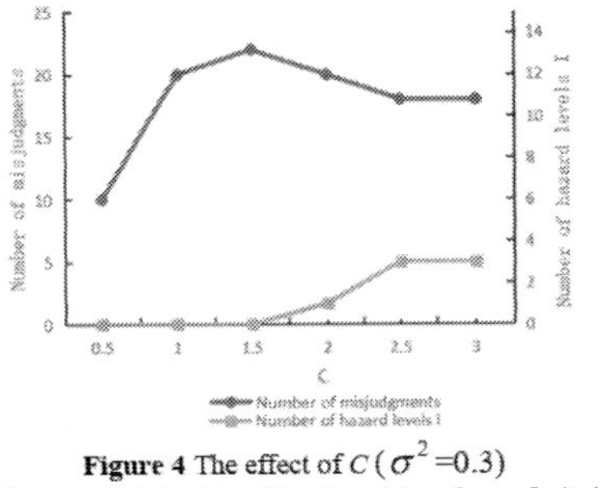

Figure 4 The effect of C (σ^2=0.3)

Using the above-mentioned RS-SVM multi-classification model to identify the risk of goaf, it is possible to comprehensively consider the risk of goaf and the nonlinear relationship between various factors, based on two kernel functions. Compared with the actual situation, the difference between the three is 13.3%. This shows that the support vector machine method, an intelligent learning method, can effectively guide the identification and evaluation of the risk of goaf in practice. It is a method of identification that is worthy of further study and popularization.

4. Conclusion

So far, the research on the risk of goaf is limited by the concealment and uncertainty of goaf. To this end, using the emerging intelligent learning method, through the learning and regular mining of limited data, seek out gobs. The non-linear relationship between hazard and various influencing factors is as follows:

(1) Hazard identification based on support vector machine has good generalization ability in small samples. For complex systems such as goaf, the result of discrimination is reasonable and effective.

(2) RS theory can effectively reduce the number of samples, improve the training speed,and at the same time reduce the redundant attributes of the samples, so that the support vector machine analysis can be more accurate.

(3) In this paper, SVM is used to classify the risk of goaf. The appropriate nuclear parameters and penalty parameters can be accurate results, and the samples also need certain diversity and representativeness.

Acknowledgments

The author(s) disclosed receipt of the following financial support for the research, authorship, and/or publication of this article: This research work was funded by the National Natural Science Foundation of China (51874112 and 51774107); the State Engineering Laboratory of Highway Maintenance Technology, Changsha University of Science and Technology (kfj170108).

References

1. H. T. Ma, N. W. Liu, Y. H. Wang, et al. Review on research status of controlling techniques for goaf disaster in metal mille[J]. Journal of Safety Science and Technology, 2014, 10(10):75-80
2. F. Q. Gong, X. B. Li, L. J. Dong, et al. Underground goaf risk evaluation based on uncertainty measurement theory[J]. Chinese Journal of Rock Mechanics and Engineering, 2008, 27(2):323-330
3. Q. M. Wang, B. G. Xu, S. H. Tang, et al. The actual state of mined-out areas in metallic and nonmetallic mines and the countermeasures for the treament of the mined-out areas in China[J]. MINING R&D, 2009, 29(4):63-68
4. Z. Q. Chen, Y. M. Wang, S. M. Chen, et al. Stability analysis of goaf based on entropy weight method and fuzzy theory[J].MINING R&D, 2019, 39(5):39-43
5. X. D. Xu, Y. J. Zhou, S. Pang. Analysis of catastrophic instability of plastic supporting system in old goaf of gypsum mine[J]. Chinese Journal of Rock Mechanics and Engineering, 2018, 37(11):2548-2555
6. Y. P. Zhang, P. Cao, H. P. Yuan, et al. Numerical simulation on stability of complicated goaf[J]. Journal of Mining & Safety Engineering, 2010, 27(2):233-238
7. G. J. Hu, T. H. Yang, Z. Q. Hu, et al. Mathews-stability-method-based multi-angle analysis and evaluation of mined-out zones stability[J]. Journal of Mining & Safety Engineering, 2017, 34(2):348-354
8. Y. J. Li, W. Q. Lin, Y. Chen. Evaluation on risk of goaf based on AGA-BP neural network[J]. Journal of Safety Science and Technology, 2015, 11(7):135-141
9. C. Xiao, H. C. Zheng, W. Z. Wu, et al. Stability evaluation of goaf in gypsum mine based on the set pair analysis[J].CHINA MINING MAGAZINE, 2014, 23(10):107-114
10. W. Wang, Z. Q. Luo, L. X. Xiong, et al. Stability evaluation of goaf based on improved matter element extension model[J]. Journal of Salety and Environment, 2015, 15(1):21-25
11. Z. Pawlak. Rough sets[J]. International Journal of Information and Computer Science, 1982, 11(5):341-356
12. S. Z. Wu, P. Z. Yan. Attribute reduction algorithm on rough set information entropy and its application[J].Computer Engineering, 2011, 37(7):56-61
13. S. F. Ding, B. J. Qi, H. Y. Tan. An overview on theory and algorithm of support vector machines[J]. Journal of University of Electronic Science and Technology of China, 2011, 40(1):2-10
14. L. X. Li, G. W. Wang. Sensitivity analysis of impact factors of goaf group stability and its application[J].MFTAI MINE, 2018, 2018(2):27-34
15. Y. Feng, X. M. Wang, A. B. Cheng, et al. Method optimization of underground goaf risk evaluation[J]. Journal of Central South University(Science and Technology), 2013, 44(7):2882-2888
16. H. Song, Y. Xue, L. G. Zhang. Research on kernel function selection simlllation based on SVM classification[J]. JISUANJI YU XIANDAIHUA, 2011, 2011(8):133-136
17. Y. G. Liu, Z. H. You, L. P. Cao. A novel and quick SVM-based multiclass classifier[J]. Pattern Recognition, 2006, 39(11):2258-2264

Comparison of M5 Model Tree and Nonlinear Autoregressive with eXogenous inputs (NARX) Neural Network for urban stormwater discharge modelling

EL GHAZOULI Khalid[1,2], *EL KHATABI Jamal*[1], *SHAHROUR Isam*[1] *and SOULHI Aziz*[3]

1 Laboratoire de Génie Civil et géo-Environnement, Université de Lille, 5900 Lille, France

2 Laboratoire d'Analyse des Systèmes, Traitement de l'Information et Management Industriel, Université Mohammed V, Rabat, Morocco

3 National Higher School of Mines, Agdal Rabat, Morocco

Abstract. This paper presents a comparative study of two data-driven modelling techniques in forecasting urban drainage stormwater discharge based on rainfall prediction. Both M5T and NARX (Nonlinear Autoregressive with eXogenous inputs) Neural Network are used for 30 minutes storm water forecasting. Data are collected from watershed area of 3315 ha, located in the city of Casablanca in Morocco. The results show that both models provide good results, but however with better performances of the NARX model.

1. Introduction

Major cities are facing urban flood issues due to urbanization and climate change. Real time control (RTC) systems allow an improvement in the performances of sewer networks through the control of the different network components and the optimization of storage within retention ponds and main collectors (Beeneken et al., 2014). The essential input parameters in RTC systems are rain and flow forecasts. Artificial intelligence techniques had shown good performances in modelling nonlinear systems with a considerable reduction in computing time compared to hydrologic and hydraulic models (Abou Rjeily et al. 2017). Two artificial intelligence techniques, namely, neural networks and M5 Model Tree were discussed in several applications such as sediment transport modeling (Bhattacharya et al. 2007), river flow forecast (Solomatine and Khada. 2003), evaporation estimation (Rahimikhoob. 2014). This paper presents an extension of this comparison to urban drainage application.

2. Methodology and material

2.1. Methododology

Artificial Neural Networks

Neural networks are commonly used in artificial intelligence techniques. They showed good performances in nonlinear system simulation and time series prediction (stock forecasts (Pang, X et al 2018) and urban flood prediction (Berkhahn et al 2019)). They are considered as black box containing the information to be learned. In the beginning, the architecture of the neural network is composed of layers and nodes without any information or knowledge of the simulated phenomenon. During the learning stage, the weights of the neurons that connect the different nodes are determined according to an optimization algorithm to minimize the error of the output of the neural network and measured data.

The Nonlinear Autoregressive with eXogenous inputs (NARX) neural network given by equation (1) is used in this research:

$$y(t) = f\left(y(t-1), y(t-2), \ldots, y(t-n_y), u(t-2), \ldots, u(t-n_u)\right) \quad (1)$$

where y(t): output time series, u(t): input time series, ny and nu: are the time delays representing the phenomenon dynamic behavior.

Two types of architectures of NARX are generally used. The first one is efficient in forecasting a time series value one-time step ahead and it's called series-parallel or open loop. The second one is called parallel architecture or closed-loop and it is efficient for multistep-ahead prediction.

M5T

The M5 model is based on the principle of information theory (Quinlan 1992, Solomatine and Khada. 2003). It can handle nonlinear problem by splitting it into multiple linear ones. The M5T is similar to conventional decision trees, but instead of predicting classes through a classifier it predicts continuous variables through linear regression functions at the leaves. The construction of the M5 model tree is done in two stages (Solomatine and Xue. 2004, Rahimikhoob et al. 2013). First, the M5 model tree is constructed by a divide-and-conquer method that splits the multidimensional data space into several subspaces based on a splitting criterion (Figure 1) that treats the standard deviation of the class values that reach a node as a measure of the error at that node and computes the expected reduction in the error by testing each attribute at that node . The splitting operation stops when the values of the instances reaches a node with slight variation or just few instances remain (Goyal and Ojha 2014)
The formula used to calculate the standard deviation reduction is given by (Solomatine and Khada. 2003, Pal and Deswal, 2009. Goyal and Ojha. 2014, Gyanendra et al 2016):

$$SDR = sd(T) - \sum \frac{|T_i|}{|T|} sd(T_i) \quad (2)$$

where T is the set of examples that reaches the node; Ti is the subset of examples that have the ith outcome of the potential set; and sd is the standard deviation

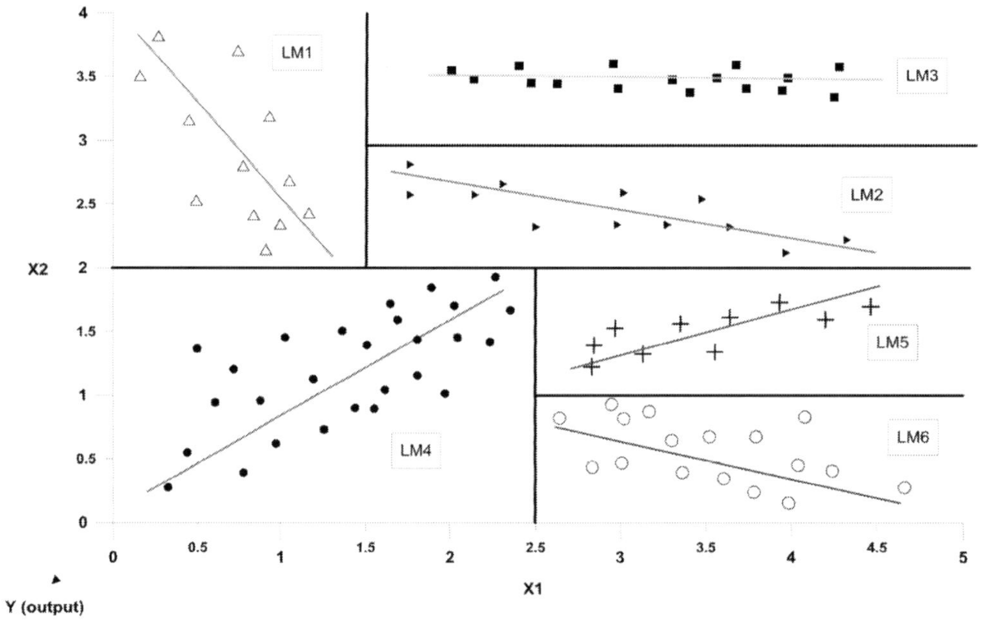

Figure 1: Splitting the input space (X1×X2) by M5 model tree algorithm

The result of the splitting is a tree with splitting rules at the nodes with expert linear models (LM) at the leaves (Figure 2). The combination of linear models can be seen as committee machine (Haykin. 1999).

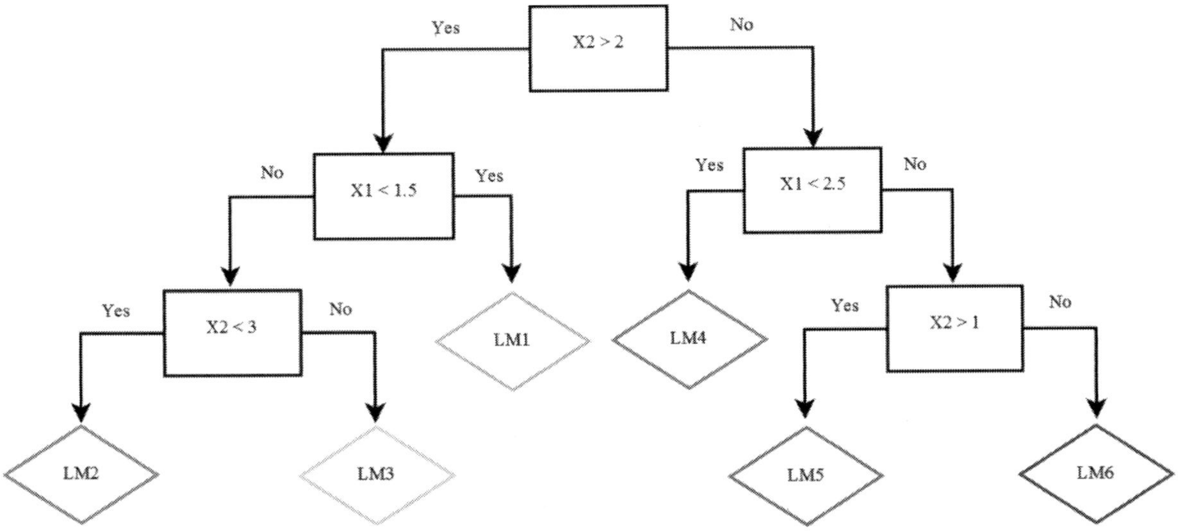

Figure 2: Diagram of the M5 model tree with six linear regression models at the leaves

The generated tree is often large and difficult to analyze and may lead to overfitting. In the second stage a pruning operation is needed to overcome this problem by replacing a subset with a leaf. In addition to this, the smoothing is performed to compensate the discontinuities that can occur between the adjacent linear models at the pruned leaves of the three.

2.2. Site description

The study area concerns a watershed of 3315 ha located in the city of Casablanca in Morocco.
The urban drainage system (UDS) in this area is composed of 485 km of both combined and separate networks (Figure 3).
The UDS is equipped with a depth meter and flow meter the outlet of the watershed. The measurements for UDS are recorded at 15 minutes time interval. A rain gauge is located in the watershed to record the rain intensity at a 5 minutes time interval.

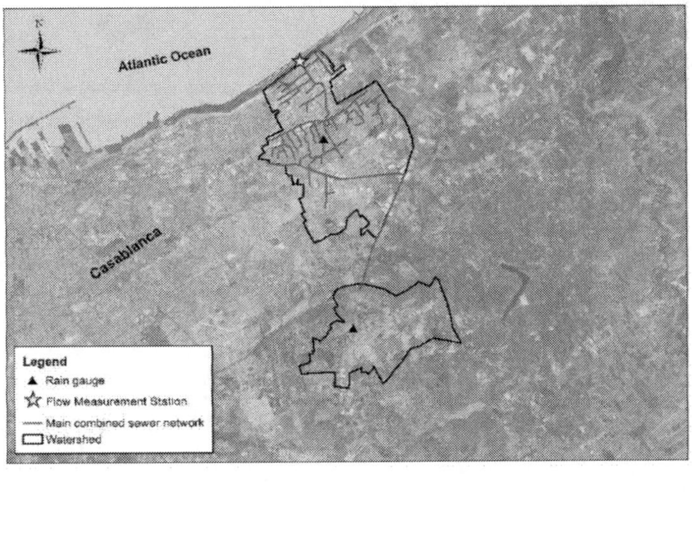

Figure 3: Area used in this study, Casablanca

2.3. Data collection

Data had been extracted from the database for a period of 4 years (January 2015 – January 2019). Since the main sewer system at the outlet of the watershed is a combined one dry weather flow had been removed to keep only data related to rain flows.
The rain dataset used in this present work is characterized by return periods summarized in table 1, which shows a dominance of low to medium rains in the dataset.

Return Period	Percentage of the sample
1 year	58%
2 year	18%
5 year	14%
10 year	2%
20 year	4%
50 year	4%

Table 1: Statistics of return periods (RP) of the rain dataset

3. Application

Both M5T and NARX were used to predict the stormwater flow at 30 minutes. Data used in the construction of the two models are the appropriate data that contribute in the stormwater flow Qt_{+30min}.
These Data are the previous rainfall data (Ret_{-80min}, …, Ret_{-5min}, Ret) and the previous measured discharge (Qt-10min, Qt-5min and Qt). Both models were trained on a data sample of 4 years.
In order to evaluate the performances of both models in predicting Qt_{+30min} for different meteorological conditions, a sample of 4 rain events with various intensities and different return period was used.
These rainfall return periods used for testing the models are: 50YRP (Year Return Period), 2 YRP, 5YRP and 5YRP.

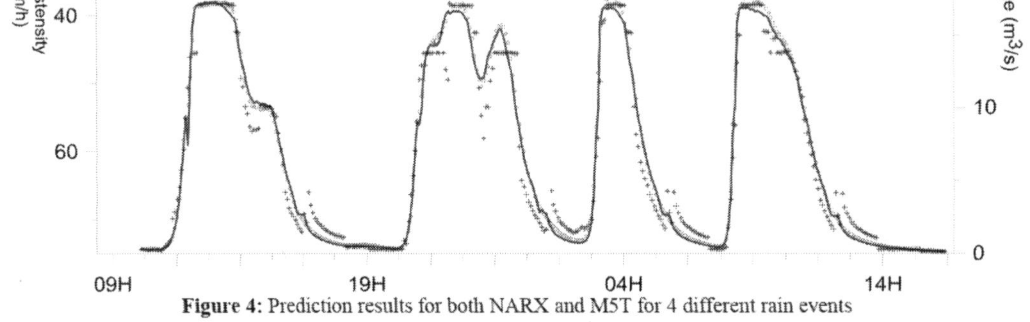

Figure 4: Prediction results for both NARX and M5T for 4 different rain events

	NARX	M5T
RMSE (m3 s)	0.726	1.330
NSE	0.988	0.959

Table 2: Performance statistics of NARX and M5 model tree

The efficiency of the models was examined through results and measurements comparison with a Nash Sutcliffe Efficiency (NSE) and Root Mean Square Error (RMSE) calculation.
Results in table 2, show a good performance of both models with a small RMSE and a high NSE. Figure 4 shows that the NARX model represents better the flow hydrograph with a slight overestimation of the peak flow (about 10%) compared to M5T, which gives better results in predicting peak flows.

4. Conclusion

This paper presented the use of two artificial intelligence techniques, namely NARX and M5T for forecasting urban drainage stormwater discharge using rainfall predictions. Both of these models could be easily implemented for practical use. They were tested and compared on data collected from watershed area of 3315 ha, located in the city of Casablanca. Results show that both models provide good results. The NARX model shows good performances in modelling all the ranges of flow with slight overestimation of peak values compared to M5T Model, which provides good results in peak flows forecasting.

References

1. Y. Abou Rjeily, O. Abbas, M. Sadek, I. Shahrour and F. Hage Chehade (2017) Flood forecasting within urban drainage systems using NARX neural network , Water Science & Technology
2. Bhattacharya, B., Price, R. K., and Solomatine, D. P. (2007). "Machine learning approach to modeling sediment transport." J. Hydraul. Eng., 133(4), 440–450
3. Beeneken, T., Erbe, V., Messmer, A., Reder, C., Rohlfing, R.,Scheer, M., Schuetze, M., Schumacher, B., Weilandt, M. & Weyand, M. (2014). Real time control (RTC) of urban drainage systems – A discussion of the additional efforts compared to conventionally operated systems. UrbanWater Journal 10
4. Berkhahn S, Fuchs, L., Neuweiler I. An ensemble neural network model for real-time prediction of urban floods.(2019) Journal of hydrology- Volume 575, ~~Volume 575,~~ Pages 743-754
5. Goyal MK, Ojha CSP (2014) Evaluation of rule and decision tree induction algorithms for generating climate change scenarios for temperature and pan evaporation on a Lake Basin. ASCE J Hydrol Eng 10.1061/(ASCE)HE. 1943–5584.0000615
6. Singh G, Sachdeva SN, Pal M (2016) M5 model tree based predictive modeling of road accidents on non-urban sections of highways in India. Accid Anal Prev 96:108–117
7. Haykin, S. _1999_. Neural networks: A comprehensive foundation,Prentice-Hall, Englewood Cliffs, N.J.
8. Pal M, Deswal S (2009) M5 model tree based modelling of reference evapotranspiration. Hydrol Process 23:1437–1443
9. Pang, X.; Zhou, Y.; Wang, P.; Lin, W.; Chang, V. (2018). An innovative neural network approach for stock market prediction. J. Supercomput, 1–21.
10. Quinlan, J.R., (1992). Learning with continuous classes. In: Proceedings of ustralianJoint Conference on Artificial Intelligence, World Scientific Press: Singapore,pp. 343–348.
11. Rahimikhoob, A. (2014). Comparison between M5 model tree and neural networks for estimating reference evapotranspiration in an arid environment. Water Resour. Manage. 28, 657–669.
12. Solomatine, D. P. & Dulal, K. N. (2003) Model trees as an alternative to neural networks in rainfall–runoff modelling. Hydrological Sciences Journal. 48(3), 399–411.
13. Solomatine DP, Xue Y (2004) M5 model trees compared to neural networks: application to flood forecasting in the upper reach of the Huai River in China. J Hydr Engrg 9(6):491–501

A Semi-Automatic Approach in GIS for 3D Modeling and Visualization of Utility Networks: Application for Sewer & Stormwater networks

Fadi Chaaban[1*], *Hanan* Darwishe[2], and *Jamal* El Khattabi [3]

[1]Dep. of Topographic Eng., Faculty of Civil Engineering, Tishreen University, Lattakia, Syria.
[2]Dep. of Topographical Eng., Faculty of Civil Engineering, AlBaath University, Homs, Syria.
[3]Civil Engineering and Geo-Environment Laboratory (LGCgE), Lille University, France.

Abstract. This paper presents a semi-automatic methodology proposed for 3D modeling of utility networks in GIS environment. The ModelBuilder in ArcGIS (ESRI) software is used for implementing this methodology, by developing two tools to automate the construction processes of 3D networks. The first presents a tool to create a 3D Manhole layer from points layer, and the second is a tool to create a 3D pipe layer. For both tools, a work algorithm has been built, in addition to designing user interfaces elements. These tools are stored in a Toolbox called "3D Manhole & Pipe.tbx". The two previous tools were tested and applied to spatial data for a proposed residential area. The final 3D model of the residential area includes the sewage and stormwater networks, as well as other spatial data such as buildings, parks, roads, etc. This model is able to spot the intersection points in the network, visually or using the 3D analysis available in the software, allowing us to identify problems to be processed and resolved before starting a project, leading consequently to time and cost savings, effort and money. The proposed methodology is an easy and an effective way to build 3D network models (sewer, water..etc), and the developed tools allow the implementation of a set of necessary processes needed to build 3D networks.

Keywords: GIS, 3D Utility Networks, 3D Modeling, ArcGIS, ModelBuilder.

* Corresponding author : fadi.chaaban@tishreen.edu.sy

1 Introduction

In general, several researches have studied 3D modeling of cities in the GIS environment [1, 2]. The development of 3D urban models, both above and underground, allows for new approaches to urban planning and management [3, 4]. These approaches serve in emergencies, disaster planning, urbanization verification, and infrastructure networks.

Geographic Information Systems (GIS) have a great ability to represent, display, query and manage spatial data (2D or 3D). It also has the ability to manage utility networks such as sewerage, Stormwater networks, water distribution, telephone networks… etc. The representation of 2D networks is related to various obstacles, such as overlapping lines that represent different networks, and various vertical elements (manholes). Moreover, 3D analysis is not possible using 2D geometry. Therefore, attention to the 3D representation of these infrastructure networks has increased rapidly in recent times [4].

Three-dimensional modeling of utility networks is an important mean of networks design, implementation, management and maintenance. During the modeling process, a wide range of processes and procedures was faced to reach the final model. This research proposes a semi-automatic methodology for 3D modeling of infrastructure networks in GIS environment, by developing two tools to automate the construction processes of 3D networks.

The provided methodology in this research contributes to building network models in an easy and an efficient way that saves time and effort, and ensures a correct way to build the spatial and topological relationships between different components of the networks. This methodology includes the following objectives: (1) Building a Geodatabase for infrastructure networks of a small residential area, (2) Development of two tools to automate the construction processes of 3D networks in the model builder environment (ArcGIS platform), (3) Export the model and all its elements to a WebScene interface to make it easily available to users, and to exploit the data associated with it.

2 Material and Methods

2.1 Study area

The study area is located in east of Tartous city centre (Syria) (about 3 km), between longitude 35.9236 °E to 35.93029 ° E, and latitude 34.8976 °N, 34.8901 °N (Fig. 1). It covers an approximately area of 16 hectares. It is a proposed residential area by [1]. The residential area consists of 54 residential blocks. It includes also commercial area (with a floor area of 3600 m^2, gardens, playgrounds and water bodies. Geodatabase was built by dividing the space into a group of layers: buildings, road network, malls, schools, kindergartens, parks, playgrounds, water bodies, etc. based on WGS 84 / UTM zone36N projected coordinate system.

2.2 Work platform and software

Essentially, this research is based on the ArcGIS platform developed by ESRI [5], ArcGIS is a platform to create, manage, share, and analyse spatial data [6]. It allows representing spatial data in 2D and 3D. It has been used in many of global scientific researches such as [7, 8].

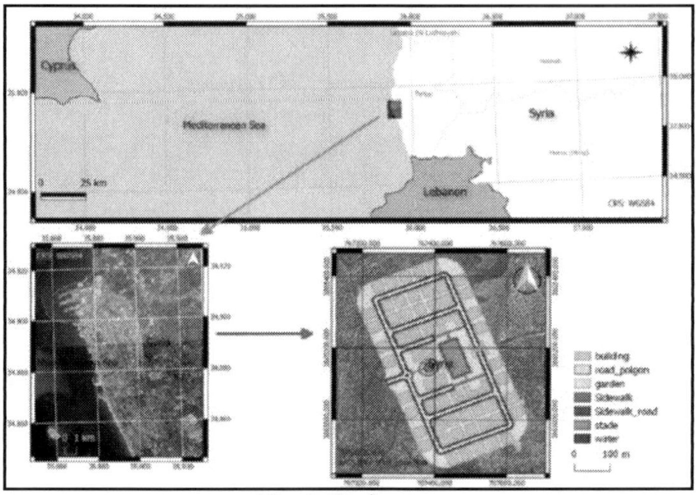

Fig. 1. Study area

2.3 ModelBuilder and geoprocessing in ArcGIS

2.3.1 Geoprocessing

Geoprocessing generally depends on the systematic implementation of process Series that are applied to spatial data to create new layers (new derived spatial data). ArcGIS consists of two environments for completing geoprocessing, namely ArcToolbox and ModelBuilder.

2.3.2 ModelBuilder

ModelBuilder is a visual programming language to create, modify, and manage models within ArcGIS. It is used to automate workflow using the geoprocessing tools available within ArcGIS (ArcToolbox) [9]. There are two basic modes for creating ModelBuilder. The first mode allows exploratory work to be performed. The other one is an application where generic tools can be built, reused and shared. In 2012, [7] designed two general models using ModelBuilder that can be used for geoprocessing linear features (calculate changes between two linear elements). Many researchers have also used ModelBuilder in global scientific research such as: [8] to calculate the change in night lights in China, and [10] who implemented a tool for facilitating the process of urban regional risk assessment.

2.4 Preparing spatial data for 3d modeling

2.4.1 Creating layers

The necessary feature classes of modeling (the sewer and Stormwater networks layers) were created within ArcCatalog, where each feature class was grouped into its own feature dataset (Fig. 2). These layers are: network pipes (line feature class) and network manholes (point feature class). Fig.2 shows the attribute tables of each feature class. line feature class includes the following fields: ID, Name, Diameter (mm), BtmPipeStart (m), MIdPipeStart (m), BtmPipeEnd(m), slope_Thousand, Depth_Start (m), Depth_End (m), DEM_Design (m), Shape_Length (m). Point feature class includes the following fields: ID, Name, elvation_designe (m), input1 (m), input2 (m), output (m), Depth (m).

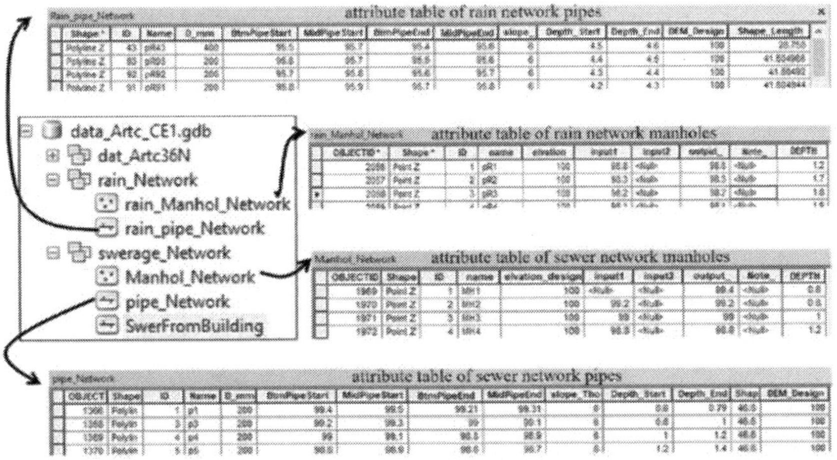

Fig. 2. Spatial database and attribute tables of networks pipes and manholes

To verify the data quality, a new topology was created for both networks. some topology rules of ArcGIS were used such as: Must Not Overlap, Must Not Intersect, Must Not Have Dangles, Must Not Self-Overlap, Must Not Self-Intersect, Must Be Covered By Endpoint Of, Must Be Single Part, Endpoint Must Be Covered By. The errors were corrected or excepted. Fig. 3 shows the 2D corrected spatial data (pipes and manholes) for both networks (sewer and rain) after applying the symbology.

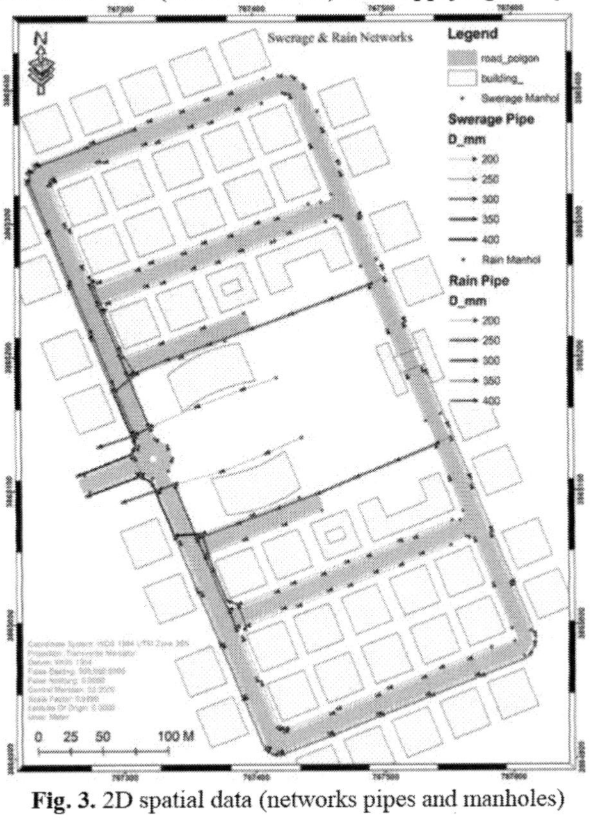

Fig. 3. 2D spatial data (networks pipes and manholes)

3 Results and discussion

3.1 Automating the construction processes of 3D networks

In this stage, two tools were developed (in ModelBuilder) to automate the construction processes of 3D networks from 2D layers. The first presents a tool to create a 3D Manhole layer from points, and the second is a tool to create a 3D pipe layer from lines. For both tools, a work algorithm has been built (Fig.4 a, b), in addition to designing user interfaces elements (Fig.4 c, d). These tools are stored in a Toolbox called "3D Manhole & Pipe.tbx".

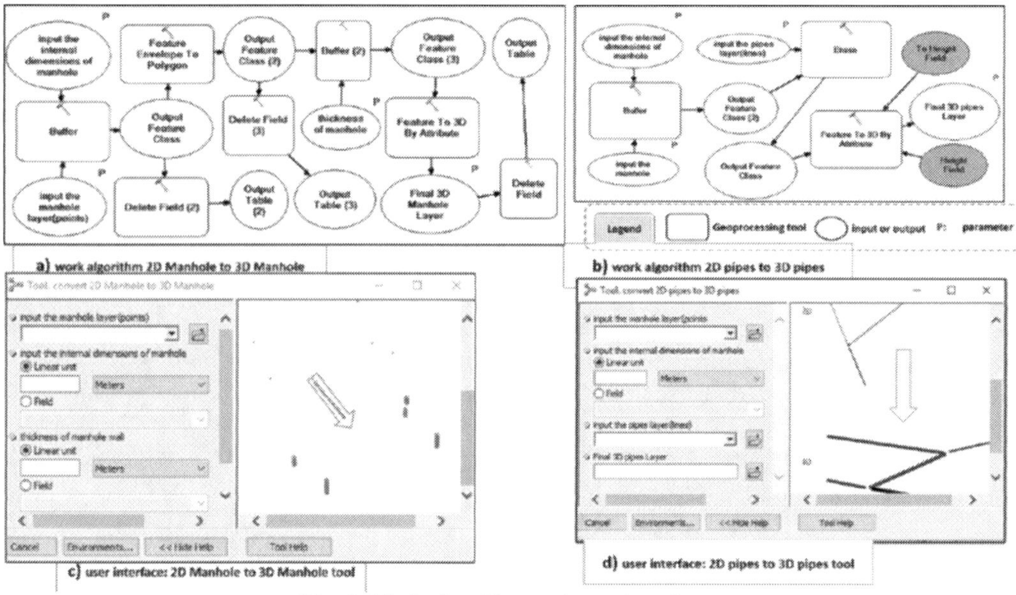

Fig. 4. Work algorithm and user interfaces

Fig.4 (a) presents a work algorithm for converting 2D manhole to 3D. It consists of the following tools: Buffer, Feature Envelope to Polygon, Delete Field, Feature To 3D By Attribute and the Outputs of each tool. Fig.4 (b) presents a work algorithm for creating a 3D pipe layer from lines. It consists of the following tools: Buffer, Erase, Delete Field, Feature To 3D By Attribute and the Outputs of each tool.

3.2 Test and apply the developed tools

The two previous developed tools were tested and applied to the spatial data presented previously (in 2.4 paragraph), which contain the required and necessary fields. Firstly, the 2D Manhole layers were converted to 3D Manhole layers, then the 2D pipes layers were converted to 3D pipes layers. In the end, three-dimensional layers (x, y, z) were obtained where the cartographic representation and visualization are possible in an environment that reads this type of 3D data (such as ArcScene). Fig. 5 represents the final 3D model of the residential area including the sewer and Stormwater networks (Manholes & Pipes), as well as other spatial data such as buildings, gardens, roads, etc. Through this model, This model is able to spot the intersection points in the network, visually or using the 3D analysis available in the software, allowing us to identify problems to be processed and resolved before starting a project, leading consequently to time and cost savings, effort and money.

The final stage of the modeling process is to share the project with other users (using ArcGIS online) by exporting the 3D model with all its elements to a WebScene interface to make it readily available to users, and to exploit the data associated with it. This interface allows to search for any feature, as well as to display the attribute data associated with any selected feature. A set of selected scenes and created bookmarks can also been exported, that can be viewed as a video within a web page.

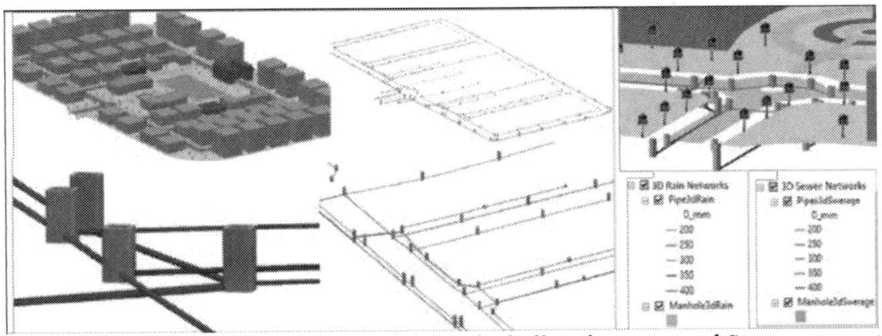

Fig. 5. Final 3D model of the residential area including the sewer and Stormwater networks

4 Conclusion

In this research, a semi-automatic methodology was proposed for 3D modeling of infrastructure networks in GIS environment, by developing two tools in order to automate the construction processes of 3D networks. A range of results was obtained: (i) The use of GIS reduces the effort to prepare spatial and attribute data for digital models, and is widely used as a tool to assist analysis, planning, management and decision-making. (ii)The ModelBuilder is an easy-to-use application to accomplish multiple and complex work tasks. (iii) The methodology presented is flexible and adaptable according to the status and type of Utility Networks. (iv)The 3d model can be shared with other users as WebScene which can be readily and exploitable.

References

1. H. Darwishe, F. Chaaban, Procedural 3D modeling of cities in Geographic Information Systems environment using ESRI CityEngine (published in Arabic). Albaath univ. magazine. **39**,11(2017)
2. K. N. Edvardsson, *3D GIS modelling using ESRI's CityEngine: a case study from the University Jaume I in Castellon de la Plana Spain* (Doctoral dissertation), (2013).
3. J. Benner, A. Geiger, K. Leinemann, F. et al., Flexible Generation of Semantic 3D Building Models, Work. Next Gener. 3D City Model, **49**, pp. 17–22.(2005)
4. T. Becker, C. Nagel, And T. H. Kolbe. *Semantic 3D modeling of multi-utility networks in cities for analysis and 3D visualization*. in Lecture Notes in Geoinformation and Cartography, pp. 41–62 (2013)
5. ESRI, "Environmental Systems Research Institute,". [Online]. Available: https://www.esri.com. [Accessed: 01-Apr-2019].
6. ArcGIS for Developers. [Online]. Available: https://developers.arcgis.com. [Accessed: 01-Apr-2019].
7. F. Chaaban, H. Darwishe, Y. Battiau-Queney, B. Louche, E. Masson, J. El Khattabi, E. Carlier. Using ArcGIS Modelbuilder and Aerial Photographs to Measure Coastline Retreat and Advance: North of France," J. Coast.Res.,**285**, pp.1567–1579(2012)
8. S. Scheider And A. Ballatore. Semantic typing of linked geoprocessing workflows, International Journal of Digital Earth, pp. 1–9.(2017)
9. D. W. Allen.*Getting to Know ArcGIS ModelBuilder*(Redlands,CA:Esri Press,2011)
10. M. Zhao And X. Liu. Regional risk assessment for urban major hazards based on GIS geoprocessing to improve public safety, Saf. Sci. **87**, pp. 18–24 (2016)

Standardized system for evaluating groundwater pollution

Khemmal Hichem Yakoub[1], *Hani Azzedine*[2]

1 laboratoire : Unités de recherche des matériaux émergeants, université Ferhat Abbas Sétif, Algérie.

Abstract. This paper presents a methodology that will permit the groundwater pollution potential of any hydrogeologic setting to be systematically evaluated. Pollution potential is a combination of hydrogeologic factors, anthropogenic influences and source of contamination in any given area. This methodology includes only the hydrogeologic factors which influence pollution potential.

1. Introduction

Groundwater has a major contribution to agricultural, industrial, drinking, and other municipal uses. The preparation of aquifer vulnerability maps has become crucial in recent years.

The DRASTIC method is a familiar method developed in the US Environmental Protection Agency (USEPA) by Aller et al. (1987) and this method has been applied in several regions by different researchers (Cameron and Peloso 2001; Al-Adamat et al. 2003; Baalousha 2006; Jamrah et al. 2007; Erhan S, et al. 2009; Ramaraju A, et al. 2019). It is the most commonly used overlay and index method all over the world, in Quebec 2018 this method was recommended by the ministry of the environment and fight against climate change to delineate wellhead protection areas.

Our objective is to achieve vulnerability assessment of groundwater using DRASTIC Method and the GIS techniques in the region of Setif which is located in the northeast of Algeria. DRASTIC layers were created during the preparation of the vulnerability map, using parameters such as groundwater level, recharge, aquifer environment, topography, and hydraulic conductivity. The aquifer vulnerability map was prepared by overlapping the layers using GIS.

2. Method and materials

2.1 Drastic method:

The name DRASTIC is taken from the initial letters of the seven parameters used to evaluate the intrinsic vulnerability of aquifer systems. The following symbols are used in the computation of DRASTIC vulnerability index.

A numeric value between 1 and 5 reflects the degree of influence of each of these parameters or their weight in the calculation of the indices. So the depth to water and the nature of the vadose zone exert the most critical influence, followed by recharge. Conversely, the topography is the parameter with the least impact, followed by the type of soil. Finally, the nature of the aquifer and its conductivity have a moderate weight.

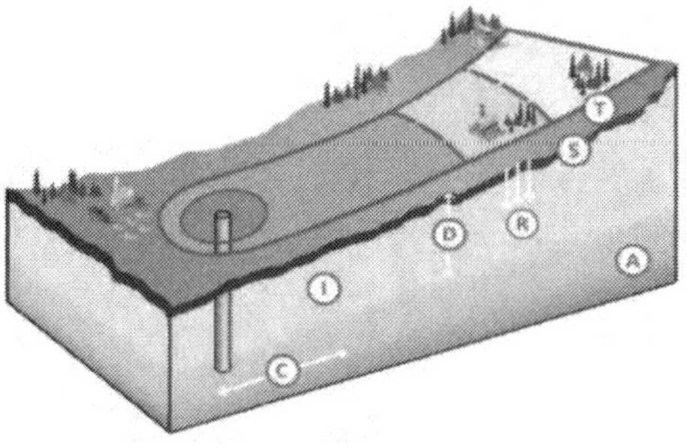

Parameters	Weight
Depth to water	5
Recharge	4
Aquifer media	3
Soil media	2
Topography slope	1
Impact of vadose zone	5
Conductivity of the aquifer	3

Figure 1: The DRASTIC method, its parameters and their weight

The determination of the DRASTIC index number is done by multiplying each parameter rating by its weight and adding together. Each parameter is rated on a scale from 1 to 10, a rating of 10 indicating a high pollution potential of the parameter.

1. **Table 1**: the standard rating proposed by the DRASTIC method (Guide to conducting vulnerability analyzes of drinking water sources in Quebec)

Rating	D	R	A	S	T	I	C
	Depth to water (m)	Recharge (mm/year)	Aquifer media	Soil media	Topography (%)	Impact of vados zone	Conductivity (m/day)
1	+31	0-50		Clay-Limestone-Marl	+18	Confident layer(1)	0.04- 4
2	23-31		Massive shale(1-3)	Black land		Clay	4-12
3	15-23	50-100	Igneous or metmorphic rocks(2-5)	Clay loam	12-18	Clay or silt(2-6) schist(2-5)	
4			Altered igneous or metmorphic rocks(3-5)	Silitous loam		Igneous or metmorphic rocks(2-8)	12-29
5	9-15		Tillite(4-6)	Loam	6-12		
6		100-180	Limestone, shale and sandstone beds (5-9). Solid sandstone (4-9). Solid limestone(4-9)	Sandy loam		Limestone(2-7) Sandstone(4-8) Sand and gravel with silt and clay(4-8)	29-41
7	4,5-9			Cracked clay			
8		180-250	Sand and gravel(4-9)	Peat		Sand and gravel(6-9)	41-82
9	1,5-4,5	+250	Basalt(2-10)	Sand	2-6	Basalt(2-10)	
10	0-1,5		Karst(9-10)	Thin soil or gravel	0-2	Karst(8-10)	+82

N.B: each material is classified according to the standard rating proposed by the drastic method. it also proposes a range of quotation indicated between () in this table.

The DRASTIC Index is then computed by applying a linear combination of all factors according to the following equation:
DRASTIC Index = DrDw + RrRw + ArAw + SrSw+ TrTw + IrIw + CrCw

The DRASTIC index can be further divided into four categories: low, moderate, high, and very high according to the Natural Breaks classification (Jenks):

• Low (<100)
• Medium (100-140)
• High (> 140).

Each category reflects an aquifer's inherent capacity to become contaminated. The higher DRASTIC index number shows the greater relative pollution potential risk to one another. DRASTIC index is relative and dimensionless that depends on the geological and hydrogeological characteristics of an aquifer.

2.2 GIS

A geographic information system is a computer tool for representing and analyzing all the elements that exist on earth and all the events that occur there. A working methodology that consists in representing the collected information in the form of thematic maps, was carried out using Arcgis.

3. Results

3.1 Depth to water:

The observation of the water depth map reveals three different depth classes from:
- 15m to 23m around Kaaboub and Amusement Park.
- 23m to 31m around Ain Sfia
- + 31 m in the rest of the area.

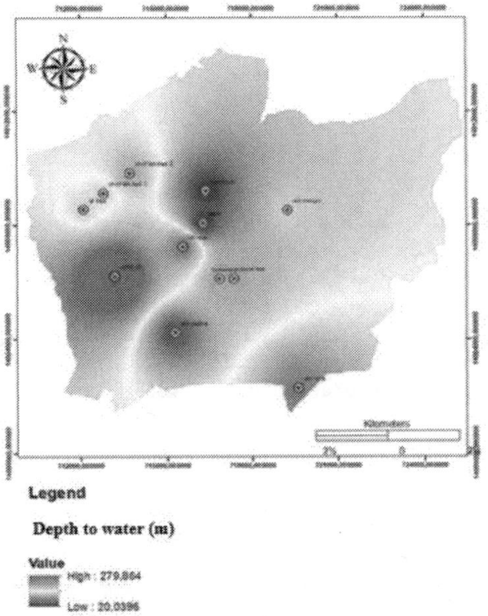

Figure 2: Depth to water rating map

Table 2: Dimensioning of the water depth parameter

Range (m)	Dr	Dw	Dr×Dw
15-23	3	5	15
23-31	2	5	10
+31	1	5	5

3.2 Recharge:

Estimating regional recharge, or at the aquifer scale, is often sufficient for the purposes of calculating DRASTIC indices. Only one value can therefore be generally used (The Ministry of Sustainable Development, Environment, and Fight Against Climate Change: Guide to conducting vulnerability analyzes of drinking water sources in Quebec (2018)). Mapping the recharge area is the second step in identifying areas that may be affected by pollutants. The largest recharge areas are the most intolerable to contamination. The calculation leads to a value of 115mm / year. The DRASTIC rating gives a rating of 6 for the range which is between 100 mm / year and 180 mm /year. We chose 4 because the calculated value is near to the range which has a rating of 3. To do not inflate the recharge rating we chose the value of 4. we would choose the rating of 6 if it was ≤150mm/year.

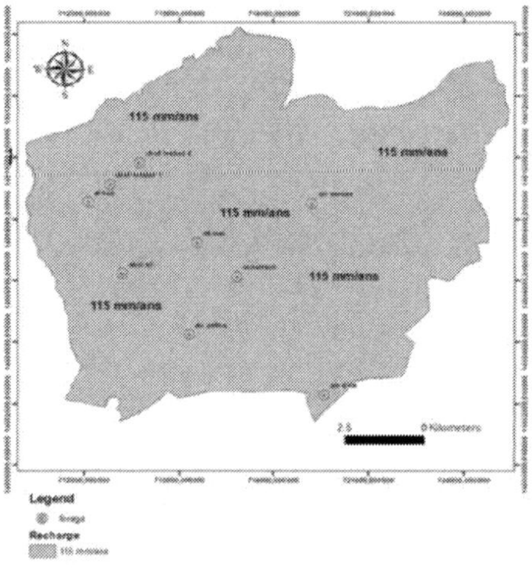

Figure 3: recharge rating map

Table 3: Dimensioning of the recharge parameter

Range	Rr	Rw	Rr×Rw
115 (mm/year)	4	4	12

3.3 Aquifer media:

The typical rating can be used if a little knowledge is available about the hydraulic properties of the formation. Otherwise, a choice in the range of the ratings must be made. For rock aquifers, it is usually the fracturing level that must be taken into account. For granular aquifers, it is the granulometry and the homogeneity of the sediment which must be taken into account. It's obvious that permeable formations have a higher rating score than low permeability and impermeable formations. The DRASTIC method gives a range from 9 to 10 to karst formations, we chose the rating 9 because of the presence of marl in the fractured karst.

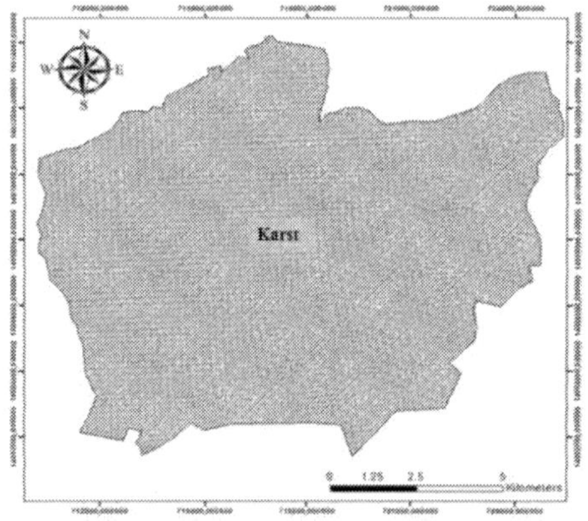

Figure 4: Aquifer media rating map

Table 4: Dimensioning of the aquifer media parameter

Formation	Ar	Aw	Ar×Aw
Karst	9	3	27

3.4 Soil media:

Throughout our study area, the surface of the ground by its impermeable nature plays a role of protector against the infiltration and the evolution of the pollutant on the contrary in the south-west part the surface of the ground is characterized by the Sandy clay that is a little permeable. The drastic method gives a rating of 1 to clay and 9 to sand we chose the rating 4 for this formation.

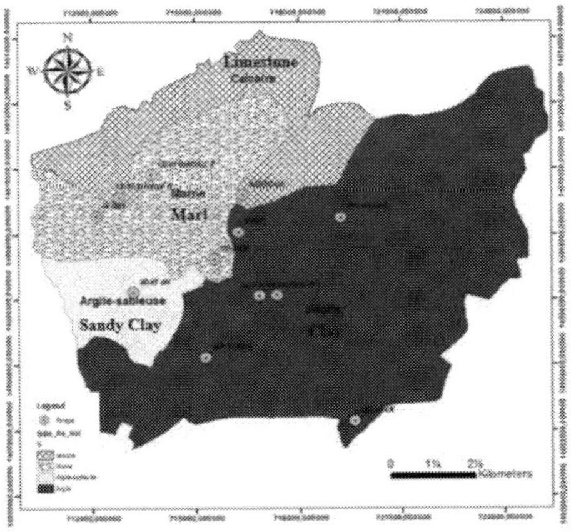

Figure 5: Soil media rating map

Table 5: Dimensioning of the soil media parameter

Formation	Sr	Sw	Sr×Sw
Limestone	1	2	2
Marl	1	2	2
Sandy Clay	4	2	8
Clay	1	2	2

3.5 Topography:

It has an influence on the infiltration and therefore the spread of the pollutant. The delimitation of the zone is based on the observation of a numerical model of elevation. Three classes were observed:

- From 6% to 12%: in the southern part of the town of Sétif. The slope does not have a great value in this area which is a part of high plains Setifian
- From 12% to 18%: in the center of the town of Sétif. It's the area that separate between high mountainous and high plains.
- +18%: in the northwestern part of the town of Sétif which is known as a mountainous area. The slope in this area has a significant value.

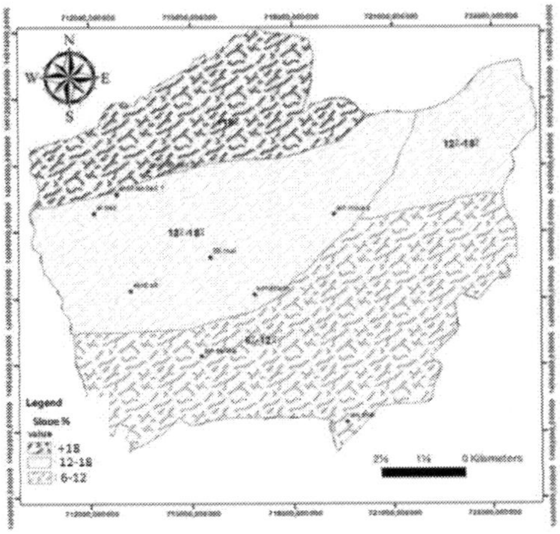

Figure 6: Slope rating map

Table 6: Dimensioning of the topography parameter

Range(%)	Tr	Tw	Tr×Tw
6-12	5	1	5
12-18	3	1	3
+18	1	1	1

3.6 Impact of vadose zone:

The same conditions concerning the standard values of the parameter on aquifer media are applied.
The rating range for clay formations varies between 2 and 6 according to the formation which is included with it and its permeability.

- We chose the rating 4 for limestone because of its low fracking.
- The marl is 50% limestone and 50% clay which gives a rating of 4 for this formation.

- We chose the rating 3 for the formation marl-limestone because of the nature of limestone in the central south of Setif which is hard.

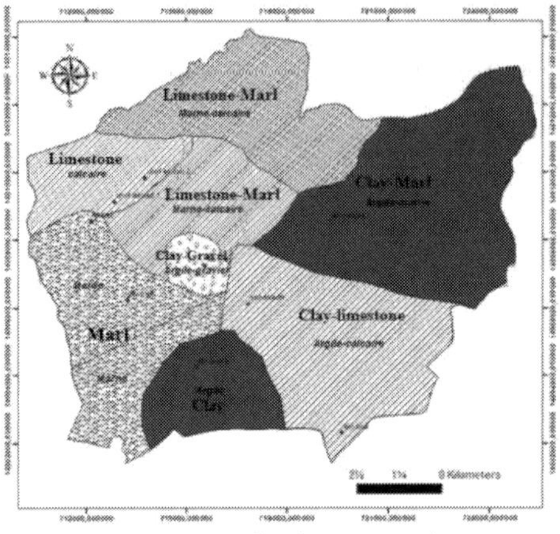

Figure 7: Impact of vadose zone rating map

Table 7: Dimensioning of impact of vadose zone parameter

Formation	Ir	Iw	Ir×Iw
Clay	2	5	10
Clay'n Limestone	4	5	20
Marl	4	5	20
Clay'n Gravel	6	5	30
Clay'n Marl	3	5	15
Limestone	4	5	20
Marl-Limestone	3	5	15

3.7 Conductivity:

The conductivity of the aquifer is between 0.04 m / d and 14 m / d. The observation of the conductivity map shows two classes:
• from 0.04 m/d to 4m/d in the center of the town of Sétif.
• From 4 to 12m/day away from the center, the permeability increases.

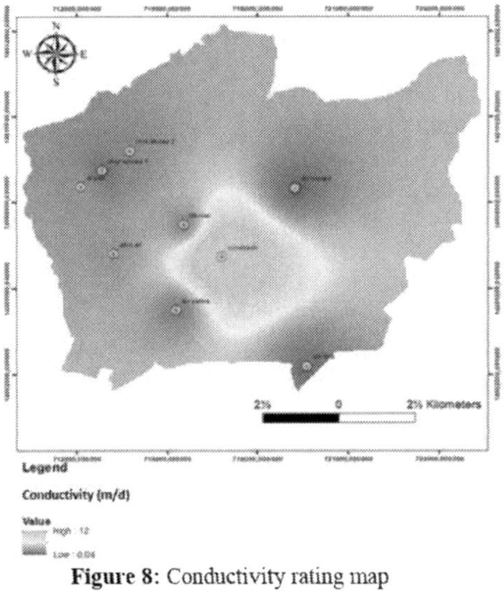

Figure 8: Conductivity rating map

Table 8: Dimensioning of conductivity parameter

Range (m/d)	Cr	Cw	Cr×Cw
0.04 - 4	1	3	3
4 – 12	2	3	6

3.8 DRASTIC index map:

The following figure illustrates DRASTIC INDEX MAP in the commune of Setif. Each of DRASTIC parameters has been expressed as thematic layer using ArcGIS. All generated maps were used to assess intrinsic groundwater vulnerability to pollution.
Only one zone of vulnerability was determined in the commune of Setif.
According to the DRASTIC scores in the vulnerability map 100% of the study area have a low potential for contamination which is between 68 and 89.
The highest value of the vulnerability is located in the center of the commune because of the low depth of the water table and the nature of the vadose zone which is permeable.

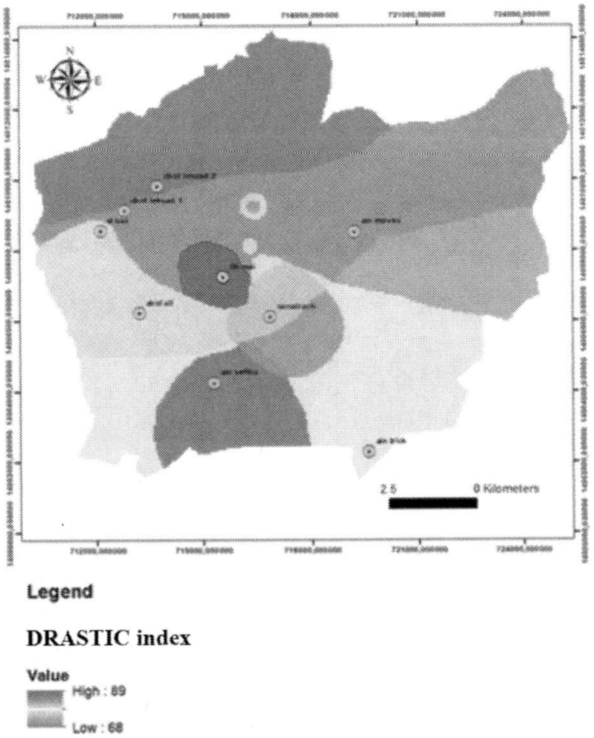

Figure 9: DRASTIC INDEX MAP in the commune of Setif

4. Conclusion

This method was prepared to assist planners, managers and administrators in the task of evaluating the relative vulnerability of areas to groundwater contamination from various sources of pollution. Once this evaluation is complete, it can be used to help direct resources and land use activities to the appropriate areas. This methodology includes also the creation of a system which can be readily displayed on map.
It is obvious that all the drastic parameters interact, dependent variable. Their selection is based on quantitatively developed and rigorously applied data coupled with a subjective understanding of real-world conditions in a given area.
Finally, we present an effective and economical protocol for assessing groundwater vulnerability and thus planning human activities.

References

1. Aller L., Benett T., Lehr J.H. and Petty R.J., (1987). DRASTIC: a standardized system for evaluating ground water pollution potential using hydrogeologic settings. Office of Research and Development, Environmental Protection Agency USA.
2. The Ministry of Sustainable Development, Environment, and Fight Against Climate Change (2018). Guide to conducting vulnerability analyzes of drinking water sources in Quebec

Application and Development of Smart Mine in China

Ying Xu[1, *] , *Junqiang* Wei[2],*Xiekang* Zhou[2],*Feng* Sun[2]

[1]State Key Laboratory of Nickel and Cobalt Resources Comprehensive Utilization, Jinchuan, 737100, China

[2]School of Mines, China University of Mining and Technology, Xuzhou , 221116, China

Abstract. The development of coal mining from ancient times to today has experienced the artificial stage, the mechanized mining stage, and the automated mining stage. Now it is developing towards a safer, greener and more efficient unmanned and smart mine. This article suggests that the development of smart mine is a process of continuous improvement. The overall goal of the smart mine is divided into three development stages, from the smart working face and driving face (1.0 era) to the smart of the whole mine (2.0 era), and finally to the subversion of smart mining technology (3.0 era), the state and coal enterprises have given great attention. At present, there are more than 40 mining areas to carry out smart mine construction plans, but there are still many unresolved problems; Finally, it gives a prospect for the development and realization of Smart Mine.
Keywords: Smart Mine; Unmanned mining; Development path

1 Introduction

The rapid development of science and technology, a new generation of research such as big data, high performance computing, artificial smart calculation and other information technology is becoming more widely, promoting geological information science major step forward towards a more smart goal , Smart Planet[1] concept came into being. Smart Mine is a part of the construction of Smart Planet. Smart Mine[2] is an unmanned mine that actively senses, automatically analyzes, and quickly processes towards the production, occupational health and safety, and technology. The essence of Smart Mine is safe mines, high-efficiency mines, and clean mines; Mines including coal mine, metal mine and non-metal mine. This paper takes coal mine as an example to illustrate.

Coal mining has experienced four stages from ancient times to present era.It is artificial mine stage, mechanized mine stage, automated mine stage and Smart Mine stage. In the artificial stage, people mainly carry out mining activities by manual and simple tool drafting and digging, and the production mainly depends on manpower, and the efficiency is extremely low; In the mechanical mine stage, the manual operation is replaced by mechanized production. The mines use a large amount of mechanical equipment and

* Corresponding author: xuying@cumt.edu.cn

blasting construction in mining, transportation, upgrading and production assistance. The number of mine workers is greatly reduced. The safety production technology support capability and the level of safety production of coal mine enterprises is greatly improved; In the stage of automated mining, coal minc enterprises comprehensively digitize comprehensive information such as mine geography, geology, mine construction, mine production, mine safety, coal mine transportation et al., so as to realize the improvement and integration of integrated automation systems, and carry out the initial construction of management informationization and engineering digitization, to achieve automated mines; Smart Mine is based on the research of mechanical mine and automated mine, awareness-based mining research on coal mining enterprises through integrated automation, information management and engineering digital of mastery, combined with Internet + [3], artificial smart, big data [4], virtual simulation, visualization, spatial information technology, network integration and cloud technology, 3D simulation and virtual reality and other advanced information technologies in an efficient, safe, green mining as the goal, the construction of highly smart, automated, user-friendly smart of mine, to create a " unmanned mining "mode of production as the ultimate goal, to realize Smart Mine [5-7]. Over the past decade, coal mining technology and equipment developed rapidly, significantly increasing mechanization and automation, providing a solid foundation for the smart exploitation. But now the majority of our information technology companies in the mechanized mine stage and automated mine stage, only a small number of coal mining companies are into stage of Smart Mine.

The construction of Smart Mine is to improve the efficiency of mine production and operation, fundamentally ensure the safe production of coal mine enterprises, promote the integration of the coal mine enterprise upper and lower systems, and strengthen the management system of coal mine enterprises, avoiding the unclear responsibility of safety accidents and ensuring the security of system data transmission realizes the sharing and information sharing of production and operation of coal mine enterprises, so that the data generated by each subsystem plays the most important role. Reduce the occupation of personnel, achieve unattended operation in a dangerous environment, establish a new mine production management mode, and achieve high efficiency and high safety performance for coal mining enterprises. With the continuous deepening of the reform of coal enterprises, the level of informatization of coal mine enterprises is also getting higher and higher. Smart Mine can realize the application of smart information technology, so that the entire system of mines has independent thinking, reaction and execution capabilities, and realizes comprehensive information of human-machines. Integration and responsiveness minimizes human factors and provides support for business managers in production management. In the end, the mine will achieve safe, green and efficient mining, and achieve concept innovation, system innovation, management innovation and technological innovation.

2 Smart Mine Development Path and Key Technologies under Construction

China's smart coal unmanned mining technology has entered the smart working face and driving face stage since 2010 (1.0 era), and is currently in the process of smart of the whole mine (2.0 era), and will eventually enter the smart of mining technology subversion. (3.0 era), to achieve the goal of safe and efficient coal unmanned mining.

2.1 Smart working face and driving face stage——1.0 era

In the smart working face and driving face (1.0 era), the smart mining mode of "smart control + remote intervention" was put forward, and the development of fully mechanized mining equipment, smart mining, accurate positioning and navigation, accurate detection of geological information et al. Building smart remote operation technology platform to realize smart mine working face. The following are the key technologies for Smart mine 1.0:

Smart mining face system: Smart fully mechanized mining face control system, realize automatic control of coal mining operation and remote control; Improve the technology of coal cutting machine memory cutting, hydraulic bracket with machine automation, automatic face straightening et al. Establish monitoring center along the slot or ground to realize on-site/concentrated/remote three-level network management of the equipment. The system can automatically control the production capacity of the shearer according to the load of the transportation system; realize the coal mining of mobile equipment through wireless network coverage. The reliable communication under the working environment, the combination of video, voice and data, solves the communication bottleneck of the whole working face, and automatically tracks and monitors the video of the coal cutting process;

Smart rapid driving system: Smart coal roadway, rock roadway, smart rapid driving, drilling and bolting combined support technology, smart advanced detection technology, roadway driving smart control and other technologies to solve the problem of alternate mining and driving, to achieve the lack of people and even unmanned mining face.

Downhole precision positioning navigation system: The technical basis of the unmanned working face is high-speed underground communication system and high-precision underground positioning and navigation system, which can accurately collect information such as safety, production and management required at any place and at any time, and ensure the timeliness of information. The software platform of the equipment positioning system such as the unmanned working face shearer is formed by the software platform of high-precision electronic map and the secondary development technology of GIS.

Accurate detection of geological information system: According to the three-dimensional contour of the working face environment obtained by laser scanning, a dynamic three-dimensional modeling technique is used to establish a dynamic three-dimensional model of the working face environment. It is necessary to always grasp the real-time position, attitude and working conditions of all kinds of fully mechanized mining equipment under the working environment, in order to give an operable mining guidance under the comprehensive analysis of the overall working environment. It is required to be able to fuse the position and posture of the equipment, the working condition information and the three-dimensional model of the fully mechanized mining equipment to obtain the dynamic equipment simulation model, and then unify the environment and equipment, and finally completely reproduce the state of the working face in the computer.

Hazard source smart prevention and control early warning and smart decision system: The hazard source smart early warning and disaster mitigation system is based on risk warning. Through real-time online monitoring of different types of hazard sources, the hazard source risk index evaluation algorithm is used to conduct real-time online assessment of the risk of disasters in various mine areas to determine the risk, type, rating and solution. Through the deployment of various types of sensors in the area where the energy release type and the constrained failure type hazard source are likely to occur in the underground, real-time monitoring, analysis and uploading of the hazard indicators of various hazard sources can be realized. Real-time smart analysis of the risk level of the hazard source by using the hazard source disaster mechanism, trigger condition criteria, mine operating conditions and hazard factors monitoring index values that have been mastered, judging the type and level of risk, and conducting downhole sound and light alarms, online in-situ warning.

2.2 Smart of the whole mine——2.0 era

The 2.0 era is an era of information interconnection and inter-communication of the entire mine, visualization of the entire mine, smart decision-making and automatic operation. From "the smart of a working face" to the " smart of the entire mining system", the smart of the entire mine has time and space services. The characteristics of integration and command and decision-making, and finally the transformation from the Internet of Everything to the Connection of All Things. The 2.0 era includes the following key technology systems:

Smart mine ventilation, drainage and other auxiliary production systems: Smart ventilation, transportation and other ancillary production systems provide a guarantee for the continuous and stable production of the working face.

Mine full station equipment facility health smart management system: Equipment prediction and health management technology have received extensive attention in recent years, mainly in the military, aerospace and other fields began to carry out specific research. For the large-scale electromechanical equipment of coal mines, the characteristics of some simple faults are mainly studied. The complex fault characteristics of the entire electromechanical system are still in the embryonic stage. The main process and content of health management is the management and health maintenance of the whole life cycle of large mechanical systems, including the establishment of system health indicators, inspection and testing, health evaluation and health recovery. The main methods and means for carrying out the health management of large-scale mechanical systems include various inspection methods, condition monitoring, fault diagnosis, health assessment, residual life prediction and other technical methods, as well as maintenance, maintenance, repair and transformation.

Mine ecological construction smart system: In order to minimize the impact of coal mining on the surface ecological environment, the mine ecological reengineering system, namely the ecological system reconstruction technology of pre-mining prevention and control, in-mining control, post-mining treatment, and the repeated restoration of the secondary ecosystem restoration technology . Before the coal seam mining, the surface ecological environment will be protected from wind and sand, and the ecological environment will be able to resist the coal seam mining impact, and the mine ecological environment self-protection function will be enhanced. In the process of coal mining, the green mining technology based on the adaptive mining technology of coal resources should be innovated to minimize the impact of coal resources mining on the surface ecological environment. Land reclamation is carried out after coal mining to restore and enhance the surface ecological environment. When the mining of the lower coal seam causes the influence of repeated mining on the surface ecological environment, the surface ecological environment should be repaired timely to prevent the damage of repeated mining on the surface ecology.

Smart coal mine centralized management system: Integrate research results, build a smart coal mine system based on cloud computing data center, with safety management and control platform and four-dimensional integrated management platform as the core, effectively integrate and centralize all subsystems of the mine, to guide and regulate the operation of various production systems and links, and to build a centralized management system for smart coal mines.

2.3 The subversion of smart mining technology——3.0 era

With the development of the times, the emergence of the Internet, artificial smart, cloud computing, big data, new technology, these new technologies are combined with the traditional coal mining industry, the formation of new technology, new thinking, new ideas of smart mine. The formation of Smart Mine will eventually break the restrictions of the traditional mining industry and change the mining methods of traditional coal mine, such as coal gasification, fluidized mining and direct underground coal power generation and other new mining modes, achieving unmanned and smart mining underground. Subvert the traditional mining mode, transportation mode and utilization mode of solid resource development; open up a new mining industry model, lead the technological revolution of mineral resources mining, realize the clean, efficient and eco-friendly development of deep mineral resources, and open up new space for economic growth; Deep in-situ liquefaction technology, deep in-situ gasification technology, and deep in-situ electro-chemical technology to complete the new mining mode of solid coal resources.

3 Smart Mine Construction Practice in China

With the increase in our coal mining smart technical aspects of R & D investment, to achieve a breakthrough, leading the direction of the coal industry in the international mining smart. Smart coal mining has been carried out in more than 40 mining areas, including Shendong Coal Group, Ningxia Coal Industry Group, China Coal Group, Shanxi Coal Chemical Group, Tongmei Coal Group, Jinmei Coal Group and Jizhong Energy Group. But strictly speaking, the smart working face under construction is in the stage of Smart Mine 1.0, and have not fully realized unmanned mining face. The following brief introduction to some of the smart working face representative.

Huangling Mining Group 1001 working face the first application of mechanized mining smart control system (SAM) to achieve the smart exploitation. Control system supporting the use of electro-hydraulic control system of hydraulic support (SAC type) and winning machine control system, the " unmanned operation, one-man patrol" Visualization smart remote intervention type mining, to " two people in the monitoring center of transportation lane in the working face visual remote intervention control, one person patrol in the working face" normalization run of results, monthly production of 170300 t, the annual production capacity of more than 2 million t, productivity increased by 25%, production workers decreased from 11 to 3 people, more than 7 million yuan savings in labor costs, the level of safety for a larger increase. The working site of the monitoring center of the transportation lane in the underground working face of Huangling Mine is shown in Figure 1.

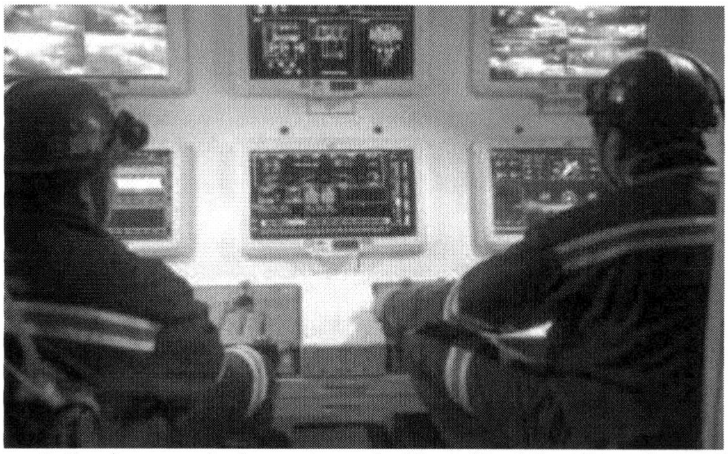

Figure. 1. Roadway monitoring center under Huangling Mine (Li et al., 2019)

Shandong Energy Zaozhuang Mining Group Binhu Coal Mine 16108 working face, by SAC bracket electro-hydraulic control system, TDECS shearer memory cutting system and working face video monitoring system were adopted ,to achieve a fully automated under the thin seam mining conditions combined with smart exploitation of remote visualization intervention. By the electro-hydraulic control system of hydraulic support, shearer cutting system memory, a video monitoring system to collect data and face images uploaded to the monitoring center located in the roadway, the operator of the work face can be achieved in the monitoring center automatic operation with the machine, shearer cutting smart memory "one-click start and stop" and the remote control of the shearer brackets and all actions, the average monthly production of coal amounted to 105 000 t, compared to non-smart working face coal produced more than 45 000 t, and the production staff of the working face was operated by the original two coal mining drivers, six bracket workers sectional machine pull frame were turned into one person for remote control, and the two people were patrolled. The annual labor cost was 1.152 million yuan, and the production level was safe. Significantly increased.

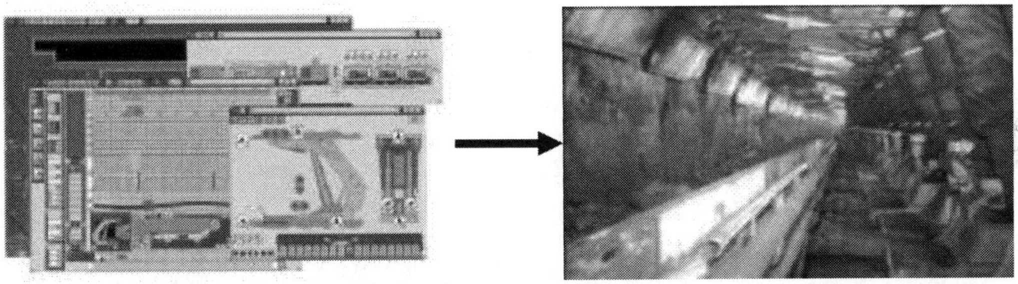

Figure. 2. Shearer memory cutting and Remote control shearer operation of Binhu Coal Mine (Wei et al., 2019)

Shendong Shangwan Coal Mine 8.8 meters of fully mechanized mining face among the highest in the world face, the face's hydraulic support, transport aircraft, transport vehicles and auxiliary transport adit et al. create more domestic and world record. The working face adopts domestic equipment, introduces digital and smart technology, and develops the travel system of high speed and high reliability heavy load shearer with 8.8m primary mining full height. In China, 3D printing technology was used for the first time to optimize the profile and structure of the sprocket wheel, and improve the machining accuracy, bearing capacity and reliability of the walking wheel. There are 128 sets of 8.8m electro-hydraulic control two-column shield hydraulic supports applied in shangwan coal mine. In addition to the highest support height in the world, this set of hydraulic supports can improve the working efficiency of Shendong Shangwan coal mine by more than 85%, reduce the cost by 30%, and increase the recovery rate of resources by 30%.

Tongxin Coal Mine adopts smart control system top coal caving face, built and successfully commissioned the first smart fully mechanized caving face in the 8202 fully mechanized caving face of the mine. To achieve a safe, efficient, low cost mining, resource recovery rate of more than 2.35%. The energy consumption per ton of coal is reduced by 5%, and the economic, safety and social benefits are very significant. The original fully mechanized mining face needs 20 workers per shift, but now only 10 people, the efficiency has been increased by 50%, reducing the labor intensity of workers. The equipment on the working face can cut coal automatically according to the program setting, and the support can be moved and sprayed automatically. The operator sitting in the monitoring center can carry out the full-automatic start-stop control on the equipment on the working face.

4 Technical Difficulties in the Construction of Smart Mine

According to the construction practice of Smart Mine in China, the level of automation and smart technology in coal mines is still low, the reliability of mining equipment needs to be enhanced, and the visualization technology in the harsh environment is urgently needed to be solved. Therefore, simulating human sensory and brain operations is a key factor in achieving smart unmanned mining. At present, some key smart technologies are still in the research stage, which makes it impossible to realize unmanned mining in some complicated working faces. In the construction process of smart mines, the following technical problems are mainly faced:

Coal rock interface identification technology and automatic height adjustment technology: To obtain the maximum recovery rate and reduce energy waste, the need to ensure cutting in the roof and floor area boundary, and to try to ensure smooth floor. Automatic navigation technology combined with geographic information system can preliminary to three-dimensional positioning of working face, but the geographic information system can not accurately obtain the dynamic Angle of coal seam and the coal and rock interface information, only through the recognition of coal and rock interface and real-time measurement of dynamic Angle of coal seam, can precisely control roller automatically adjustable high, make the cutting area and coal seam occurrence is consistent information.

Communication optimization technology: At present, the data transmission mainly adopts wire transmission and cable, but the disadvantage is that there are many fault points and it is difficult to maintain. Therefore, according to the requirements of data transmission, wireless transmission is adopted to realize the transmission and monitoring of all kinds of data.

About the high precision, high stability sensor technology: Although the sensors in use at present can realize the monitoring and collection of relevant data in the production process, there are still problems in the aspects of accuracy, transmission speed, data collection amount, anti-interference ability, seismic resistance, waterproof, dust penetration and stability, which restrict the further development of smart mines to different degrees.

The technology of information exchange between systems, fault self-diagnosis and autonomous decision making：At present smart mining technology in smart perception, independent decision-making and other smart level lower, it is unable to complete the equipment adaptive adjustment according to the geological changes of the smart fully mechanized mining face in the coal mining process. Further exploration should be made in the aspects of smart detection, smart analysis and smart control technology research, so as to improve the smart perception, autonomous adaptation and smart control ability of the equipment，as well as the need for smart equipment failure from the category of the fault, the fault location, on the way to explore and realize automatic judgement and processing of the system software.

Production continuity guarantee technology: In order to ensure the continuity of the production of fully mechanized mining face, it is necessary to carry out reasonable and orderly monitoring of each link and smartly solve some unexpected problems. Such as conveyor balance control technology, anti-collision technology, monitoring of mine pressure, scraper chain tension, conveyor joint strength, porosity, gas, fire, roof fooling et al.

5 Summary

The development of Smart Mine is a process of continuous improvement. The overall goal of the Smart Mine is divided into three development stages, from the smart of the mining working face (1.0 era) to the smart of the whole mine (2.0 era); Finally, the mining technology subversion of the smart mine (3.0 era), and ultimately achieve mine safety, green, efficient, unmanned mining, to achieve concept innovation, system innovation, management innovation, technological innovation.

The state and coal enterprises have attached great importance and strong support to the construction of Smart Mine, and carried out a large number of theoretical and practical explorations and practices; the current coal smart unmanned mining is still in the comprehensive mining automation combined with visual remote intervention the semi-automatic 1.0 era; By the end of 2018, 145 fully mechanized mining faces in more than 40 mining areas have been implemented for smart mining.

However, there are still many unresolved problems in the realization of Smart Mine; there are low conversion rates of scientific and technological achievements, practical application of heavy concept propaganda, and poor adaptability to various complex geological conditions. At present, the distance from the real realization of smart, unmanned, and normalized applications is still far away, and it is still necessary to make great efforts in management concepts, investment, and R&D team building. With the development of related basic industries such as advanced manufacturing, sensing, computing science and technology, and with the continuous improvement of technical conditions, the only way for smart mine to make the coal mine unmanned, improve efficiency and green safety will be achieved.

6 Acknowledgements

The financial support by the Opening Project of State Key Laboratory of Nickel and Cobalt Resources Comprehensive Utilization is gratefully acknowledged.

References

1. W.Xu, T.C.Guo. The Implementation of IBM "Smart Earth" Strategy and Its Impact on China [J]. China Science and Technology Forum, **34**(03): 148-153 (2014).
2. G.F.Wang, H.Wang, H.W.Ren.Scenario Goals and Development Path of Smart Coal Mine 2025[J].Journal of China Coal Society,**43**(02):295-305 (2018).
3. S.R.Ge, Z.B.Wang, S.B.Wang. Research on Key Technologies of Internet + Shearer Smart[J].Coal Science and Technology,**44**(7):1-9 (2016).
4. Y.H.Yang, B.Zhou, J.LBi. Design of Heterogeneous Data Integration Platform for Smart Mines[J]. Industry and Automation, **41**(5): 23-26 (2015).
5. G,F.Wang, D.S.Zhang. Innovative Practice and Development Prospect of Coal Smart Comprehensive Mining Technology[J]. Journal of China University of Mining & Technology, **47**(03):459-467 (2018).
6. H.P.Xie, J.H.Wang, G.FWang, H.W.Ren,. The new concept of coal revolution and the conception of coal science and technology development[J]. Journal of China Coal Society, **43**(05): 1187-1197 (2018).
7. J.D.Fan. Innovation and Development of Smart Mining Technology in Coal Mines[J]. Coal Science and Technology, **45**(09):65-71 (2017).

An Effective Approach for Land Use Management in the Palestinian territories

Azzedine Hani

Laboratoire Ressources en Eau et Développement Durable, Université Badji Mokhtar Annaba, 23000 (Algérie). haniazzedine@hotmail.com

Abstract. The current understanding of the causes of land-use change is dominated by simplifications, which in turn, underline many environment-development policies that led to unsustainable situation in Gaza City. Therefore, this research aims at establishing an integrated land-use management framework for Gaza city based on cause-effect relationship within sustainability context. It also aims at establishing a prediction model for the relationship between the most influential socio-economic, environmental and institutional indicators that had shaped the current land-use status of the city for the period 1967-2003. The methodology presented in this research work, offers opportunities to simulate the future demand of the different land-uses based upon actual land-use conditions and other determinant factors. The determinant variables of land-use changes have been identified and prioritized using statistical analysis and Artificial Neural Network (ANN). The results were compared with other statistical techniques and expert opinions. ANN prediction model helped in drawing scenarios for future development. Combinations of socio-economic, environmental and institutional variables in addition to the actual land use for the last four years are used as a basis of land-use change explanations and modeling. These pathways indicate that land-use policies and projections for the future must not only capture the population indicators as the only drivers for land-use change but also account for the specific human resource development indicators and urban-environmental conditions. This recognition requires moving beyond some of the simplifications that persist in much of the current understanding of the causes of land-use change and its driving forces. The analysis of the local expert's opinions provide evidence support the conclusion that the simple answers found in population growth, poverty and infrastructure rarely provide an adequate explanations of land-use changes. Rather, social responses follow from changing economic conditions, mediated by institutional factors are the real causes for land-use changes in Gaza.

1. Introduction

Sustainable land-use management aims at integrating socio-economic, environmental and institutional factors into one framework that leads to formulate the appropriate policies and activities. Developing indicators framework is essential tool for tracking progress towards sustainable land-use. The selection of these indicators should be based on objective criteria and well-designed conceptual framework.

The researches done until now for developing a conceptual framework for modeling sustainable land-use management are still fragmented and incomplete attempts. Most of the reviewed researched focused on investigation the tools for special modeling of urban landuse rather than analyzing the interaction between factors that effecting the changes of urban land-use. Very few scholars described the relationship between the driving forces, environmental impacts and urban expansion in one framework [1]. Therefore, for the efficient analysis of land-use change problems, it is essential to establish a conceptual integrated framework that aims at achieving sustainable land-use management. It is essential to look for the appropriate modeling approaches and tools that could give acceptable results in analysis and predicting the available data. Traditional statistical models such as multiple regression analysis, principal component analysis and factor analysis have been very successful in interpreting socio-economic activities. However, traditional statistics are criticized as being ineffective in modeling spatial and temporal data. Attempts have also been made to develop spatially explicit models of land-use and land cover change by using several tools such as satellite imagery, aerial photography, GIS, remote sensing, agent-based modeling and ANN based modeling [2].

The research aims at achieving the following objectives:

- To establish integrated land-use management framework for Gaza city based on cause-effect relationship within sustainability context.
- To establish a prediction model for the relationship between the most influential socio-economic, environmental and institutional indicators, that had shaped the current status of the land-use in the city for the period 1967- 2003. This prediction model is the basis for drawing scenarios for future development.
- To develop recommendations for sustainable land-use management.

2. Material and Methods

Gaza City is situated in a strategic location between Asia and Africa, between the desert in the South and the Mediterranean climate in the North. The total area of the city is about 45,353 km2, which is about 12% of the total area of the Gaza Strip. The city consists of 17 neighborhoods, varying in characters, which are formed by different residents. The environmental indicators had multi-dimensional implications and characterized by continuous deterioration. The increase the amount of the generated wastewater, increased abstraction of ground water, increase the density inside the city and increase the amounts of the annual solid waste are adding more pressure on the limited natural resources.

The analysis plan followed in this research is designed to achieve two main objectives; firstly, selecting the most influential, socio-economic, environmental and institutional, indicators and secondly, modeling the relationship between these indicators and land-use changes. Several steps have been taken to achieve these objectives using different tools (Figure 1).

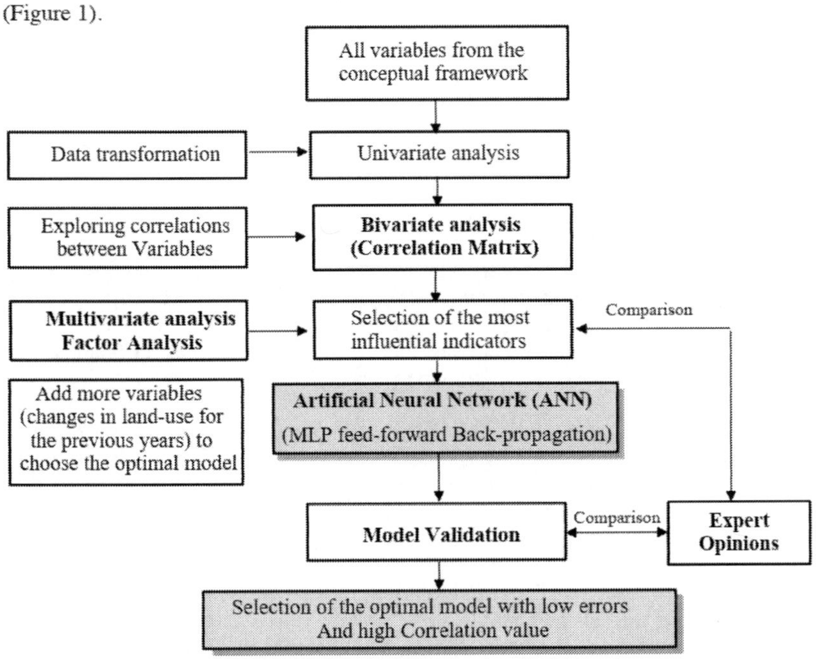

Fig. 1. Analysis Plan and Methodology.

3. Results

3.1. Statistical Analysis

Univariate, bivariate and multivariate techniques have been used to investigate how socioeconomic, environmental and institutional indicators are related to explain the land-use changes. Despite the fact that these statistical methods have several assumptions such as normal distribution of the data and linear relationships with the dependent variables, it was helpful in identifying the determinant indicators and factors that influence the changes in land-use of Gaza. The statistical analysis resulted with 10 indicates that have significant correlation with the different land-uses in Gaza and contributed by major role in shaping the land-uses for the last 37 years. These correlations vary from 0.7 to 0.95 in both directions (positive or negative correlations). These indicators are Log population under 19 years (P19), Log population density (PD), Log households connected to the wastewater (HWW), Log labour forces (LF), unemployment rate (UE), poverty rate (PR), households used cesspits for wastewater disposal (WWC), groundwater abstraction (GWA), groundwater salinity (TDS), and funds for development (FUND). Among these 10 indicators LF, PD and PR have strong correlation with all land-uses. Therefore these three indicators should be considered as the main driving forces behind land-use changes in Gaza.

The correlation coefficient (r) was calculated for each pair of variables to avoid using highly correlated variables (i.e. coefficient of correlation over 0.80) and to reduce the effect of muti-collinearity [3].

The correlation matrix reveals that population density has significant correlations among all the 10 determinate indicators. This confirms the fact that the densely areas are associated with more socio-economic and environmental problems. Therefore population density is considered as the main driving force for land-use changes in Gaza. Also groundwater abstraction has correlated significantly with other indicators, while funds for development has a moderate to low correlations with other indicators.

Factor Analysis: the main applications of factor analytic techniques are to reduce the number of the independent variables and to detect structure in the relationships between these variables, that is to classify variables. Applying factor analysis to the data on Gaza city, with the 10 indicators resulted by using minimum eigenvalue method, it is noticed that only 2 factors were successfully extracted with eigenvalue more than 1.0 (Table 1).

Table 1. Eigenvalues using principle component extraction method.

Factor No.	Eigenvalue	% of the Total variances	Cumulative Eigenvalue	Cumulative %
1	6.98	69.82	6.98	69.82
2	2.25	22.48	9.23	92.30

From the second column *(Eigenvalue)* of the table above, we find the variance on the new factors that were successively extracted. In the third column, these values are expressed as a percent of the total variance. As we can see, factor 1 accounts for 69.8 percent of the variance which means that Factor one explained more that 72 percent of the actual changes in land-use of Gaza city, factor 2 accounts around 22.5 percent. The first two eigenvalues cumulated around 92 percent of the total variance, while the other 8 factors explained less than 8 percent of the changes in the land-use areas.
Considering the first column of loading in the factor loadings matrix, shown in Table 1, indicates that this factor has a strong influence on the six indicators which are Log population density, Log households connected to wastewater network, Log households uses cesspits wastewater disposal, Log Groundwater abstraction, Log domestic water salinity and funds for development. It combined the social, environmental pressure and environmental impacts indicators. This factor is clearly identified the quality and quantity of the Gaza urban environment. Therefore it could be named as urban-environment management factor.
Considering the second column of factor loadings, it is clear that this factor has influence on four indicators and a moderate to weak influence on the remaining indicators. Since those four indicators focusing on human resource development indicators therefore this factor could be named as human resources development factor. This interpretation is consistent with the finding that this factor has a moderate effect on the other indicators due to the lower investment in the field of human resource development during the last 37 years. The slow improvement of the economical conditions in Gaza has insignificant impacts on development of social condition. These difficulties combined with the political instability in the area are hindering the efforts by the local institutions to minimize the pollution pressures nor they can provide remediation for the environmental negative impacts that threatening the public health.
Comparison with Expert Opinions and Judgments: the comparison of the statistical results with expert's opinions has significant value in evaluating the current understanding of local experts to the influential indicators that is driving land-use changes in Gaza. Therefore 23 local experts were identified as sources for individual indicator-ranking information. Experts were chosen according to their knowledge in the field of land-use management as well as their professional ability to provide ranking for the identified socio-economic, environmental and institutional indicators. Each expert was asked to rank these indicators (30 indicators) in order of its relative importance and relevance to the land-use management in Gaza City. The results of expert's opinions were accomplished through the circulation of the indicators-based questionnaire.
The results of the questionnaire were averaged within each indicator, and a preliminary ranking was computed. The results shows that the overall agreement between expert's and the statistical results was only 47%. The agreements were found in 14 indicators mainly population density, population under 19 years, unemployment rate, poverty rate and funds for development. In the other hand this analysis revealed that the environmental indicators were less estimated which resulted significant differences from the statistical analysis.
Therefore the statistical analysis compared to expert's opinions provide evidence support the conclusion that the simple answers found in population growth, poverty and infrastructure rarely provide an adequate understanding of land-use changes. Rather, individual and social responses follow from changing economic conditions, mediated by institutional factors.
Opportunities and constrains for new land-use are created by markets policies, increasing influence by regional factors. Various urban-environment conditions react to and reshape of the impacts of drivers differently, leading to specific pathways of land-use change. It is precisely these combinations that need to be conceptualized and used as the basis of land change explanations and modeling. This conclusion makes it more complicated to the traditional statistical models to produce an accurate prediction model for the changes of land-uses that could deal with complex relations between the indicators and its underlying factors. In addition, the collected data on the selected indicators showed non-linear relationships and non normal distribution.
Therefore, the next stage of modeling requires advanced techniques that provides a nonlinear function mapping of a set of input variables into the corresponding network output, without the requirement of having to specify the actual mathematical form of the relation between the input and output variables, it should has the versatility for modeling a wide range of complex non-linear phenomena.

3.2. Analysis and Modeling Using Artificial Neural Network (ANN)

ANN used here as analysis and predictor tool aims at exploring and modeling the relationship between the predictor variables (the determinant indicators, the inputs) and predicted variable (land-use indicators, the output), even if the relationship is very complex and not easy to articulate in the usual method of "correlations" or "differences between groups".

The advantages of the artificial neural network ANN approach are: its ability to handle nonlinear functions, to perform model-free function estimation, to learn from data relationships that are not otherwise known and, to generalize to unseen situations.

The output variables are fixed for all trained models, which are repressing the percentages of land-uses, divided into five sub-indicators which are residential (RES), roads (ROD), industrial (IND), recreational (REC) and agricultural (AGR) indicators. The output data have been divided into three subsets; training (60%), testing (20%) and validation (20%). The results of these models, hereafter, called land-use models.

A total of 20 combinations of input variables presented in four groups are investigated.

The first group consists of nine models with a combination of the SEEI(t) variables and additional land use variables from the previous years LU(t-1), LU(t-2), ..., LU(t-7).

The second group has four models with combination of SEEI variables from the previous 4 years {i.e. SEEI(t), SEEI(t-1),.... SEEI(t-4)}.

The third group has three models with combination of additional variables from land-use variables LU(t-1), LU(t-2), LU(t-3)and SEEI(t), SEEI(t-1), SSEI(t-2) and SSEI(t-3) variables.

The fourth group consists of five models with combination of SEEI(t), in addition to the changes in land use expressed by ΔLU(t). Additional ΔLU(t-1), ..., ΔLU(t-5).

The optimal ANN model's structure that resulted in minimum error and maximum efficiency during both training and testing was selected for validation. This experiment was done for developing MLPs; then, based on the performance of various trained MLPs with different input combinations, the best network were selected.

The data for each input combination were divided into a training set, a validation set, and a testing set. The training set consists of a set of examples used only for learning, (i.e. to fit the weights of the network). The validation set is a set used to adjust the network parameters such as network architecture (for example number of hidden layers and neurons, or number of training cycles). The test set is a set of examples used only to assess the generalization performance of a trained neural network, (external valuation). Hence, MLP was applied to the prediction of land-use areas in four phases: (1) design of the network and of inputs from historical data; (2) network training using a subset of inputs; (3) testing of the neural network using the full data set of the inputs; and (4) using the information from the neural network to forecast changes. To avoid over-training of the network, the neural network was trained with a different set of data by changing the trained set of data in random order. Therefore about 60% of the investigated data set was chosen as training data set (i.e. 21 cases), while other 20% (i.e. 8 cases) chosen as cross-validation data set and the rest cases 20% (i.e. 8 cases) were chosen as testing data set. This procedure has been repeated several times with different data inputs. In order to select the optimal model, about 16 models have been produced. Each model used different data sets have a significant influence on the predicted flow.

ANN Model used successfully as an optimization tool in a number of case studies. ANN well suited to the task of selecting an appropriate combination of inputs to a model as it has the ability to search through large numbers of combinations, where interdependencies between variables may exist. The main steps in the optimization process aims at determining the input variables, which have a significant influence on the predicted land-use changes. Consequently, the followed procedure analyzed different combinations of the influential socio-economic, environmental and institutional (SEEI) indicators in addition to the antecedent land-use values (i.e. $LU_{(t-1)}$, $LU_{(t-2)}$ .... $LU_{(t-n)}$ or the SEEI indicators for the previous years (i.e. $SEEI_{(t-1)}$, $SEEI_{(t-2)}$,....... $SEEI_{(t-n)}$. Input variables including combination of the data from pervious years on both LU and SEEI are also trained.

For the network structure identification, the current study employed a procedure of time series of land-use changes. This procedure suggests training a network with minimum input variables initially (i.e. the 10 SEEI indicators with 37 data sets), and systematically increasing the input vector dimension by adding more variables from the previous year and training the new network. This procedure, though adding complexity to the model, considers the ability of additional input variable to capture any nonlinear residual dependencies.

The number of hidden neurons in the network, which is responsible for capturing the dynamic and complex relationship between various input and output variables, could be identified by various trials [4]. The trial-and-error procedure started with one hidden neurons initially, and the number of hidden neurons was increased up to 4 during the trails. The small number of hidden neurons refers to the limited number of observations. For each set of hidden neurons, the network was trained in batch mode to minimize the average-square error at the output layer. In order to check any over-fitting during training, a cross-validation was performed by keeping track of the efficiency of the fitted model. The training was stopped when there was no significant improvement in the model's efficiency, and the model was then tested for its generalization properties.

The selection process of the desired optimal network model is composed of two consecutive stages, which indicate the values of efficiency. In the first stage, statistical accuracy measures such as coefficient of determination (R-square), Average Squared Error (ASE) and the Mean Absolute Relative Error (MARE) on both training and testing data sets to filter out the most promising optimal networks. The values of both ASE and MARE close to zero indicate a better performing model. The values of R^2 range from 0 to 1, with higher values close to 1 indicating better model performance. Statistically, an optimal network is defined as the one with the best overall accuracy measures. In the second stage, the predicted and actual outputs graphical responses for both training and testing sets for the selected most promising networks. Based on the overall graphical evaluation of each model's performance, the absolute optimal network can easily selected.

Group 1 : The model performance continues to increase significantly with the additional antecedent land-use information to the SEEI(t) variables. Note that a combination of time-series data on land-uses and SEEI(t) variables in the input vector significantly improve the performance among the inputs models in the first group. The performance of model (6) indicates that addition of antecedent land-use information for more than 4 year is not significant in the optimal input vector. The best performance was achieved for Model 5.

This model has efficiency of 83% to 97% for the training data sets and 91 to 97% for validation data sets.

Group 2 : Among the second group, the performances of the model 11 indicates that addition of antecedent SEEI information for more than 3 year did not significantly improve the performance of the model. Therefore, model 10 is best model among this group, however, this model has performance ranged between 71% to 83%, which is lower than the optimal model in group 1.

Group 3 : Models 13, 14 and 15 in the third group presents the combination of additional variables from both land uses and SEEI indicators from the previous years. These additional inputs did not improve the model's performance significantly.

Group 4 : Similar to the first group the model performance continues to improve for additional 4 years of changes in land use. Model (19) has better performance among the other models among this group, however, its results still lower than it found in the model number (5).

The graphical comparison of the different models' performance is given in Figure 2. It is obvious between the developed ANN model 5 has close values of R-square in both training and testing phases. This clearly indicates the powerful function approximation features of the back-propagation in the multi-layer perceptron ANN method.

The average squared error (ASE) and the mean absolute relative error (MARE) for all landuse categories have been calculated. The results confirmed the findings of the comparison between the observed and predicted percentages of land-use categories. It is obvious that the model 5 has the smallest values for both ASE and MARE.

The optimal network found was MLP with 30 input predictive variables and one hidden node. This resulted a model that has a good performance for land-uses with average performance ratio of 93% and ASE of 0.003. The input vector is represented by the 10 determinate factors and the land-use areas for the previous 4 years.

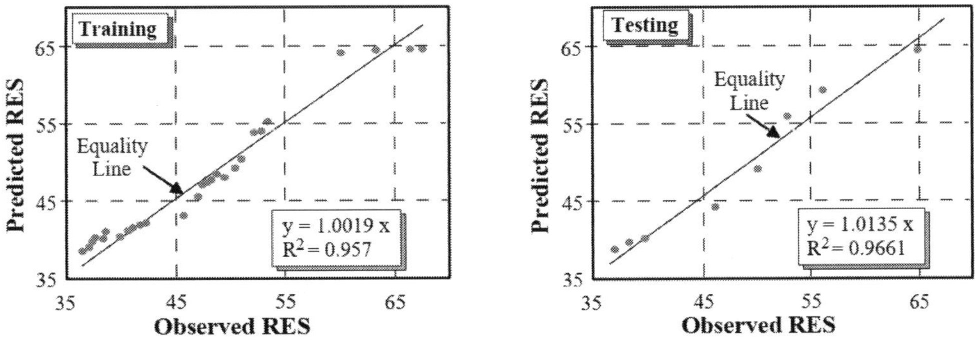

Fig. 2. Observed versus Predicted Residential Areas (Training and Testing).

The land-use percentages model can be represented by the following compact form:

$$LU(t) = ANN\begin{Bmatrix} P19_{(t)}, PD_{(t)}, HWW_{(t)}, LF_{(t)}, UE_{(t)}, PR_{(t)}, WWC_{(t)}, GWA_{(t)}, TDS_{(t)}, FUND_{(t)}, \\ LU_{(t-1)}, LU_{(t-2)}, LU_{(t-3)}, LU_{(t-4)} \end{Bmatrix}$$

Where $(LU)_t$ is the Land-use areas at time (t) which is equal to the sum of (RES + ROD + IND + REC +AGR) at time (t).

4. Conclusion

In this research, a sustainability approach for land-use management has been developed based on cause effect relationship. The sustainability approach gives insight into the management of land-use as a function of the complex interaction of a large number of socioeconomic, environmental and institutional factors. The effectiveness of such an approach depends on finding the non-traditional ways of dealing with land-use management. The methodology presented in this research work, offers opportunities to simulate the future demand of land-use changes based upon actual land-use conditions and other determinant factors.

The determinant variables of land-use changes have been identified and prioritized using multi-criteria analysis with Artificial Neural Network (ANN) compared with other statistical techniques and expert opinion. The main findings of these analysis techniques are:

• Population group under the age of 19 years old followed by population density, access to wastewater network, percentage of labour forces, unemployment rate and poverty rate are the most socio-economic driving forces for land-use change.

• Percentages of households use cesspits for wastewater discharge, and ground water abstraction are the main pressing environmental pollution sources that associated with land-use changes. The salinity of domestic water is the most representative variable of the environmental impacts of land-use changes.

• Policies, regulations and funds allocated for development are the most institutional measure that affects the land-use changes.

References

1. A. Veldkamp, P. Verburg, Elsevier Ltd. All. **72** 1–3 (2004)
2. M.S. Eila, *Thèse de Doct., de l'USTL, France.* 194 pp. (2005)
3. J.F. Mas, H. Puig, J.L. Palacio, A. Sosa Lopez, Modeling Environmental Modeling & Software **19**, 461–471 (2004)
4. H.R. Maier, and G.C. Dandy, Environmental Modelling & Software **15**: 101–124 (2000)

Analysis and interpret road traffic congestion costs in Lebanon

Alain Saroufim[1, *], *and Elie* Otayek[2]

[1]Newly Admitted at University of Illinois at Chicago, 1200 W. Harrison St,Chicago, IL 60607-7164, USA
[2]Faculty of Engineering, Holy Spirit University of Kaslik, Kaslik, Lebanon

Abstract. The cost of environmental degradation in Lebanon was estimated to be close to $485 million per year, or 2.9 percent of the Gross Domestic Product (GDP). The large negative impacts of this degradation are the pollution occurring in urban area and is mainly caused by air pollution and with an average estimate of $145 million per year and around 0.87 percent of the GDP. This high air pollution is mostly linked to the transportation sector in Lebanon. Lebanon suffers from a daily increase in road network usage, which has resulted in gridlocks, especially in the greater Beirut area. By all means, the continued lack of policies and a political will to encourage the adoption of an efficient and reliable public transportation and to discourage the ownership of private vehicles have resulted in traffic growth that is faster than the road capacity. By all means, congestion is not a passing problem; it will continue to grow in the absence of measures to reduce traffic by adopting other modes of transportation, such as buses, trains, Trams, bicycles and others etc. This study describes the various factors that affect the costs associated with congestion, and provides well-defined methods for evaluating the corresponding costs.

1 Introduction and Background

Lebanon, a small developing country in the Middle East suffers from an inadequate transportation system. By all means, the roads of Lebanon are badly maintained and poorly designed and the most exacerbated problem is that the transportation sector rely only on vehicles especially the private one which represent 88% from a total of 1.2 million automobile [1]. A study conducted in 2002 show that the ownership of vehicle is estimated to be 3 persons per car and tend to increase by 1.5% each year. Furthermore, Lebanon is one of the countries that possess a very low vehicle occupancy of 1.2 compared to average standards for computing travel time reliability which is 1.7 [2]. Based on the Ministry of Environment (MoE), the supply branch in Lebanon covers around 1.4% of the total surface area. It consists of classified and non-classified roads. 30% are classified road which means they are under the supervision of the Ministry of Public Works and Transportation (MoPWT) and the other 70% are non-classified roads under the supervision of municipalities [3].

[1] alain.saroufim@hotmail.com

Lebanon's main cities are Beirut, Tripoli, Sidon, Jounieh, Zahle and Tyre. More than 50% of the population lives in Beirut and Mount Lebanon. This inappropriate distribution leads to various problems especially in the transportation sector. Beirut's entrances are usually highly congested, with relatively little daily variation. Between Monday and Friday, traffic peaks are from 7 to 11 AM and from 5 to 7 PM, but traffic remains fairly dense throughout the day. On Sunday evenings, Beirut's entrances are very busy, while traffic in the downtown area is relatively fluid. Therefore, according to Mediterranean Environmental Technical Assistance Program (METAP) this inappropriate demographic distribution has led to highly congested areas especially greater Beirut region and led to noticeable increase in pollution in this region where most of this pollution is caused by air pollution and its indirect costs represent 0.87 percent of the GDP [4]. The increased air pollution is mostly linked to the transportation sector in this metropolitan area, where most transportation modes used are private vehicles and almost the total absence of the public transportation modes.

This paper aims to alleviate the congestion cost due to the transportation sector in Lebanon, by proposing a methodology to estimate the cost of congestion in Lebanon.

2 Material and Methods

In order to compute the congestion cost we will group the cost into three major categories:

1. Travel time estimation
2. Cost of fuel and emissions
3. Cost of air pollution.

2.1 Travel Time Estimation

The Value of Travel Time (VTT) refers to the cost of time spent on transport. The Value of Travel Time Savings (VTTS) refers to the benefits from reduced travel time costs. For travel time estimation, we will use the method developed by Litman's [5].

1. VTT is classified as an opportunity, meaning drivers can be productive instead of being time-framed on the road.
2. VTT depends on the average speeds during different congestion levels. The data shown in the Table below describe the average vehicle speeds and share of total trip (travel time). This data has been extracted from a research made by the Ministry of Environment (MoE) and United Nations Development Programme (UNDP) for Climate Change [6] respectively.

Table 1: Vehicle average speeds and share of total trip under different congestion levels

	Average vehicle speeds (Km/hr)	Share of total trip (%)
Urban peak	17.6	51%
Urban off-peak	39.1	31%
Rural	51.3	18%

3. Finally, VTT increases when the purpose of the trip becomes so serious.

The section below shows what are the parameters that affect the VTT and describe the method to calculate it.

VTT is calculated using the following formula:

$$VTT = (W \times n_w \times s_t)/\, S_a \quad (1)$$

Where n_w: Percentage wage rate

S_a: Vehicle average speed

s_t: Share of travel

W: Average hourly wage rate calculated as follow

The data for the VTT calculation is obtained from the table below:

Table 2: VTT CALCULATION

Category	Description/Circumstance of Travel	Corresponding Cost Value	Portion of Travel Time/Travel Goal
Paid	Travel by employees when being paid, travelling to meetings or between job sites	150% wage rate	5% Commercial travel
Personal High Cost	Personal travel when discomfort and frustration is experienced	50% wage rate	20% Urban peak travel/congestion
Personal Medium Cost	Personal travel when no discomfort is experienced	25% wage rate	50% No congestion

In 2007, an investigation conducted by the Central Administration of Statistics (CAS) concluded that the average wage is approximately 702,000 Lebanese Pound (LBP) [7]. Furthermore, based on International Monetary Fund (IMF) The GDP per capita for the same year was $5,937 [8], which is equivalent to an average wage of 742,096 LBP, yielding a 5.7% difference as compared to the value obtained by CAS. The adjusted average wage rate in 2010 was 688 us$ per month. In conclusion, the average hourly wage rate W was estimated to be 4.30 us$/hr/8h workday/5 working days per week.

2.2 Cost of Fuel and Emission Impacts

Recently, road transportation sector is contributing to about 20-30% of the total emissions in worldwide [9]. Based on recent report made by the International Energy Agency (IEA), more than the 60% of the oil consumption in 2008 in the world is consumed by the transportation sector. In metropolitan areas, traffic congestion is highly to occur, therefore evaluating cost emissions in these areas is essential. This part will outline in detail, how traffic congestion affects fuel consumption and incite the key players on the importance of including traffic congestion in modeling.

Fuel consumption and air pollution emissions depend on a number of factors:

- Operating speeds
- Carburant types
- Vehicle options and conditions.

2.2.1 Operating Speeds, Carburant Types and Vehicle Options

During congestion, the operating speed would be very limited. However, studies show that a car engine of 2L idling at 700 rpm would consume a lot more fuel when operating at a speed of 100 km/h. The transport sector in Lebanon occupy the second place in energy consumption since it depends totally on fossil fuels. Based on IPT Energy Center (IPTEC) the sector of transportation is responsible for over 60% of total fuel consumption where 99.2% is gasoline [10]. Finally, in Lebanon and since the cars are the most common travel mode on the roads, we will take in to consideration the following types of carburant.

Table 3: Average Price of different type of gasoline in Lebanon

Type of carburant	Price of each liter/$	Price of each liter/$
98 unleaded gasoline	0.926$	50%
95 unleaded gasoline	0.903$	50%

The table below describe the fuel consumption and emissions for each car category. However, in order to simplify our calculation, we will assume that all cars that fall victims to traffic congestion are considered to be medium cars.

Table 4: Variation of fuel consumption and CO_2 emissions in relation to vehicle categories

Car Category	Fuel Consumption (L/100km)	CO_2 Emissions (g/km)
Small	7.95	185
Medium	8.48	200
Large	9.24	218
SUV	9.66	228

2.3 Cost of Air Pollution

A study made by Chaaban et al in 2001 alert in terms of air pollution and its health impacts resulting from road transport [11]; therefore, evaluating the damage of air pollution in terms of damage cost is necessary. Given that CO_2 occupies 97.33% of GHG [12]; thus, the cost of emission and climate change is evaluated based on the Social Cost of Carbon. The Social Cost of Carbon is the amount of harm done by every ton of CO_2. In Lebanon a value of 40$/ton would be taken.

3 Results

3.1 Value of Travel Time.

Most calculation of the value of time (VTT) evaluate the behavioral value of time-more efficient in modeling-but certain approaches evaluate the economic value of time. In this paper we will compute the VTT based on the approach that identify a considerable evidence confirming significant variation in the VTT in function of the purpose of the trip, the mode of travel and the level of congestion [13].
Users' income is evaluated and shown in the following table:

Table 5: Average hourly salary in LBP

Governorate	Average hourly salary in LBP
Beirut	6531
Mount Lebanon	5090
North Lebanon	4289
Bekaa	4498
South Lebanon	4044
Weighted Average	4805= 3.20 $

As calculated in the table the weighted average income per hour is 3.2$. However, for further checking we will derive this value from the GDP per capita in Lebanon which is 6983 $ per year and by considering 5 working days per week with 8 hours of work each day we will have as average hourly salary 3.30$ /hr. The difference between Weighted Average and the GDP method is that the GDP doesn't take into consideration the income per governorate. Nevertheless, a value of 3.30$ could be taken. The table below show the final calculation of VTT.

Table 6: VTT Calculation

Trip Purpose	% of Total Trips	% Wage Rate	Average Speed (Km/hr.)	VTT $/hr.	Total Cost
Professional Work	51	150	35	$ 7.21	
Personal (Studies, shopping, bank...)	31	50	35	$ 1.46	
Leisure, tourism, other...	18	25	35	$ 0.42	
Average VTT Cost $/hr.				$ 3.03	$ 181,683,803

3.2 Cost of Fuel and Emissions

The approximate number of cars that fall victims to congestion each day in Greater Beirut Area (GBA) is 204,135 cars. On the other hand, and since during congestion a medium car engine of 2L can be almost considered as idling at 700 rpm therefore, the total fuel consumption, as based on how much CO_2 one car emits or the amount of air consumed, is estimated to be 1.2 L/hr./car. In Lebanon the average delay per trip during peak hour time is 20 minutes, therefore since nearly all users make two trips one in the morning and one afternoon so the total average delay would be 40 minutes. Congestion is known to be dynamics and related to many factors: daytime, weekends, holidays, seasons and others... However, to simplify the calculations we will assume the same degree of congestion along the year. The following table display the total cost of extra fuel consumption resulting from congestion in GBA.

Table 7: Table summarizing the total cost of extra fuel consumption resulting from congestion

Type of Carburant	Price of each liter/$	Split in %	Number of Cars Falling in Congestion	Consumption L/h during Congestion	Total Cost/ 40 minutes delay	Total Cost /year
98 unleaded Gasoline	0.926	50%	102067	1.2	$ 75,989	$ 27,736,090
95 unleaded Gasoline	0.903	50%	102067	1.2	$ 74,101	$ 27,047,181
Total amount of extra fuel consumption per day					$ 150,091	$ 54,783,272

Concerning CO_2 emission, since each one liter of burned fuel would produce a 2.35 kg of CO_2 therefore, the total emissions resulting from 40 minutes of congestion would be:

Table 8: The total cost of CO_2 emissions based on the SC-CO_2

Type of Emission	Number of Cars Falling in Congestion	Consumption L/h During Congestion	Emission in Kg/L	Total Emission in Kg	Total Cost /Year (SC-CO_2 40$/Tonne)
CO_2	204134	1.2	2.35	385690.78 Kg	$ 5,631,085

Finally, the table below summarize the sub-totals and the total of the congestion costs.

Table 9: Total Cost due to Congestion

Congestion Cost Summary	
Cost	Total
Sub-Total VOT	$ 181,683,803.44
Sub-Total Fuel	$ 54,783,272.15
Sub-Total CO_2 Emission	$ 5,631,085.382
Total Cost	$ 242,098,160.972

To sum up, the diagram below describes the proportion in % of each category cost concerning the congestion. We can easily notice that the VTT is highest indirect cost induced from the congestion.

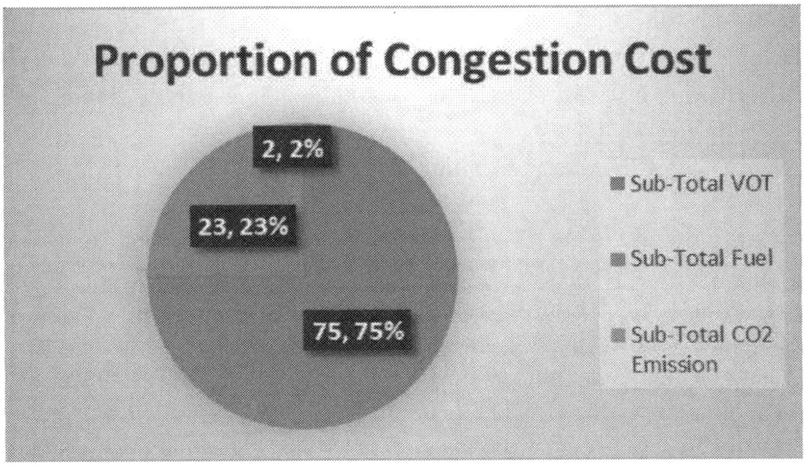

Figure 1: Proportion of congestion cost

4 Conclusion

The purpose of this study is to analyze the economic cost of road traffic congestion in Lebanon by listing a detail calculation methodology to derive the congestion cost as presented. This study suggests that road traffic congestion costs the Republic of Lebanon approximately $ 242,098,160.972 in 2017. In addition, the estimated costs of congestion are less than the true resulting costs of congestion. This low estimation is due to many reasons such as low accuracy of vehicles data in certain regions, the climate change cost due to congestion and the shift travel time impact induced by the travel mode shifting. The GDP in Lebanon is $47.54 billion therefore the total cost of congestion represents 0.509% of the GDP, which is remarkable for the congestion to attain such value in a such small country. This study incites the key players in Lebanon on how much additional indirect cost is induced by the congestion in this small country and that the problem must not be neglected.

Finally, after deriving the congestion cost, we can conclude that the problem is going to increase unless serious action is taken concerning the transportation sector in Lebanon. One of the main solutions is to encourage the public sector by making a reliable and affordable bus system, BRT, trains or by implementing policies that encourage the increase of vehicle occupancy by making HOV lanes or perhaps increase taxes on low occupancy vehicles.

References

1. C. Mansour, E. Zgheib, S. Saba, Energy Procedia-Elsevier, **6**, 261-276, (2011).
2. Federal Highway Administration (FHWA), (2018). Retrieved from: https://www.fhwa.dot.gov/tpm/guidance/avo_factors.pdf
3. MoE/LEDO, (2001). Retrieved from: http://www.moe.gov.lb/ledo/soer2001pdf/chpt5_trs.pdf
4. World Bank/ METAP, (2003). Retrieved from: http://www.moe.gov.lb/abquar/docs/refer-1.pdf
5. T. Litman, VTPI, *Techniques, Estimates and Implications. Second Edition*, (2011).
6. MoE/UNDP, Mobility Cost: A Case Study for Lebanon. Beirut, Lebanon (2015).
7. CAS, (2011). Retrieved from: http://www.cas.gov.lb/index.php?option=com_content&view=article&id=58&Itemid=40
8. IMF, Subsidy Reform in the Middle East and North Africa: Recent Progress and Challenges Ahead. IMF, Middle East and Central Asia Department, (2014).
9. A. Garcia-Castro, A. Monzon, TRB, (2014).
10. IPT Energy Center (IPTEC), (2016). Retrieved from: https://www.iptgroup.com.lb/Library/Assets/Road%20Transport%20Sector%20and%20Air%20Pollution-%20Case%20of%20Lebanon%202016-011954.pdf
11. F. B. Chaaban, I. Nuwayhid, S. Djoundoourian, Transportation Research Part D, 347-355, (2001).
12. MoE/UNDP/GEF (2016). *Lebanon's third national communication to the UNFCCC. Beirut, Lebanon*. Available at: http://climatechange.moe.gov.lb/viewfile.aspx?id=239
13. C.Fezzi, I.Bateman, S.Ferrini, Journal of Environmental Economics and Management, **67**, 58-70, (2014).

Use of Risk Analysis for Water Security Assessment

Samah Jabari[1], *Isam Shahrour*[1], *Jamal khatabi*[1]

[1]Laboratoire de Génie Civil et géo-Environnement, Université de Lille, 5900 Lille, France

Abstract. This paper presents analysis of water security using risk analysis. It contributes to an enhanced understanding of the water security issue through creating indicators and their use in the determination of the water security level. The paper presents first the methodology followed in this research, which includes three steps: identification of water risk drivers, the determination of their severity and likelihood and finally the calculation of the water security score. The paper presents the first step of the application of this methodology to Palestinian territory through the identification of the drivers of the water risk and the levels of both the severity and likelihood to be used in future water security assessment.

1 Introduction

Many countries around the world face severe water crisis [1]. For these countries, the water security constitutes a critical issue ([2] & [3]). The water security concept is large combines different factors, which are related to political, social, financial, infrastructural, environmental, managerial and natural disasters issues ([4] & [5]). This issue is becoming critical, because of rapid population as well as lack of investments in, conflicts and climate change ([5], [6], [7] and [8]). Therefore, it is necessary to conduct research to assess water security with metric expression and numerical dimensions ([7] & [9]). [2] Pointed out that quantifying water security results in an increase of our understanding of this critical issue, to establish the needs in this area and to evaluate strategies concerning water security.

Recently the risk approach was used in water security assessment [10]. It allows to identify uncertainties in this field and to characterize these risks quantitatively and qualitatively [5]. Nowadays water managers are more interested by the water risks than by maximizing the water resources performances [8]. Mason and Calow discuss the importance of risk in water security by using a set of indicators [9]. Grey and his collages identified the water security as the society acceptable water related risks [1]. The Institute for Water, Environment & Health of the United Nations University used the concept of water related risks in the definition of water security [8]. Lautze and Manthrithilake proposed water security indicators including risk management [2]. Garrick and Hall in their publication "Water Security and Society: Risks, Metrics, and Pathways" pointed out that the use of the risk concept in water security allows to provide robust diagnoses tools for the evaluation of the water security as well as its drivers [13]. At the same context other literature indicates that risk-based approach of the water insecurity improves the resilience of the water system [11].

2 Methodology

Figure 1 shows the methodology followed in this research for the use of risk assessment in water security analysis. It includes three main steps:

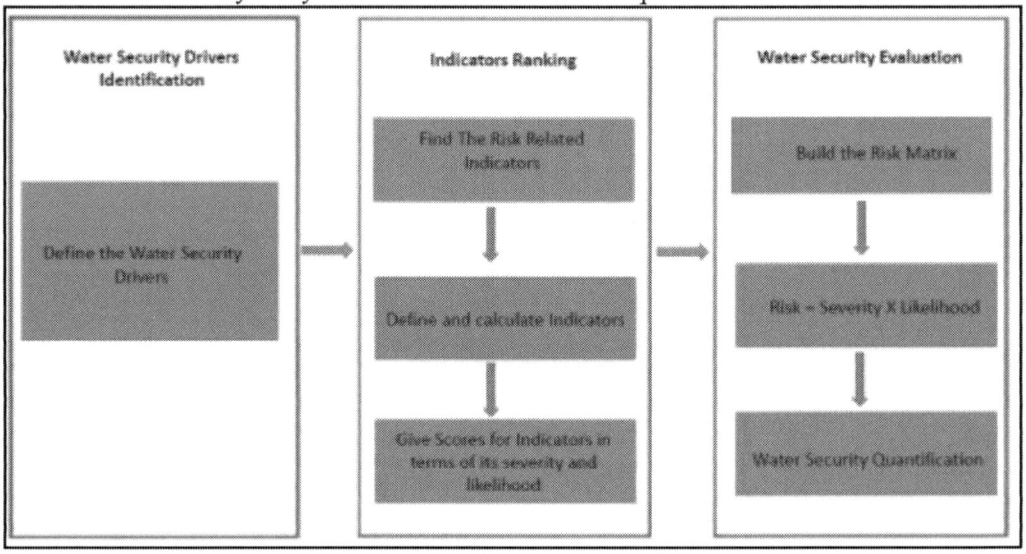

Fig. 1. Water Security Evaluation Framework

Step 1: Identification of water security drivers:
This step concerns identification of the factors (drivers) affecting water security. This step is complex, because it requires a deep and comprehensive understanding the multitude of issues related to the water security, which were discussed previously in this paper. It concerns technical, environmental, social, economic and political issues. Each indicator should be clearly identified as well as its determination and impact on water security.

Step 2: Indicators ranking:
Indicators ranking constitutes a complex and sensitive issue. It has a great impact on the analysis of water security and on priorities and strategies to be implemented in the water security plans. The use of the risk assessment approach requires the determination of the severity score and the likelihood for each driver. The former indicates the amplitude of the impact of the indicator on water security, while the latter provides an estimation of the probability of occurrence of this driver. The determination of both the score and likelihood should be based on literature review, experts consulting and benchmarking.

Step 3: Water security evaluation:
Water security is evaluated from the drivers' severity score and likelihood according to the following expression:

$$R.S = S.S \times L.S \quad (1)$$

Where;
R.S: Risk Score (from 1 to 25)
S.S: Severity Score (from 1 to 5)

L.S: Likelihood Score (from 1 to 5)

3 Application to Palestinian Territory

The Palestinian territory faces a severe state of water insecurity ([14] & [15]), because it is subjected to an increasing pressure related to climate conditions, urbanization, conflict, lack of investment and modernization in the water management [16].

Both extensive literature review and discussion with experts allowed the identification of the drivers of the water security as well as their scores and likelihoods. Drivers of the water security were classified in three categories: Water resources availability, quality of water services and governance. Table 1 summarizes the indicators associated to each category of drivers.

Water resources drivers include three indicators: Water resources availability, Annual precipitation and ratio of treated waste water. The quality of water services for the population is governed by three indicators: Water coverage, Water losses and Continuity of the water supply. Finally, for the water governance the following three indicators were identified: Roles and responsibilities, access to data and information and Stakeholders' engagement.

Discussions are under progress for the determination for each indicator both the severity score and likelihood score using for each of them the levels summarized in table 2. The result of calculation of the risk score from Eq.1 and the risk matrix expressed in table 3, then it will be used according to table 4 for the evaluation of the water security score.

The risk matrix which shown in table 3 was used to calculate the risk score for each indicator. The arithmetic mean for summation of indicator scores in a certain category will give the water security driver risk score. This score will rate the risk in to low, medium, high, and extreme level. Then risks scores and levels translated to water security scores and levels as shown in the suggesting system at table 4.

Table 1. Water Security Drivers and Associated Indicators for Palestine Territory

Security Driver	Indicators
Water resources	Water resources availability
	Annual Precipitation
	Ratio of Treated Waste Water
Quality of water Services	Water coverage
	Water losses
	Continuity of supply
Water Governance	Clear Roles and Responsibilities
	Access to Data and Information
	Stakeholders' Engagement

Table 2. Severity and Likelihood Ranking System

Likelihood Score	Likelihood level	Severity Score	Severity level
1	Rare	1	**Insignificant**
2	Unlikely	2	**Mainor**
3	possible	3	**Moderate**
4	Likely	4	**Critical**
5	Almost Certain	5	**Catastrophic**

Table 3. Risk Matrix

Risk Matrix			Severity				
			insignificant	Minor	Moderate	Major	Catastrophic
Score			1	2	3	4	5
Likelihood	Almost Certain	5	5	10	15	20	25
	Likely	4	4	8	12	16	20
	Possible	3	3	6	9	12	15
	Unlikely	2	2	4	6	8	10
	Rare	1	1	2	3	4	5
Risk Score				< 6	6-9	10-15	> 15
Risk Level				Low	Medium	High	Very high

Table 4. Water Security Scoring System

Risk Score (RS)	Risk Level (RL)	Water Security Score (WSS)	Water Security Level (WSL)
$1 \leq RS < 6$	Low	4	High
$6 \leq RS < 10$	Medium	3	Challenging
$10 \leq RS < 16$	High	2	Low
$RS \geq 16$	**Extreme**	**1**	**Very Low**

4 Conclusion

This paper presented the use of the risk assessment approach for the evaluation of the water security, which constitutes a major issue for a large number of countries. This methodology includes 3 steps: identification of the drivers of the water risk and their associated indicators, determination for each indicator of the scores of its severity and likelihood and finally the evaluation of the risk score according to Eq.1.

The paper presented also the first stage of the application of this methodology on Palestine territory. Literature review and discussion with experts allowed the identification of three categories of drivers: water resources, quality of water services and water governance. For each driver, the paper presented the associated indicators. The paper presented also the scales to be used in the risk assessment.

Work is under progress to collect data from some cities in order to conduct water security benchmarking between these cities.

References

1. United Nations Educational, Scientific and Cultural Organization. WWAP (United Nations World Water Assessment Programme)/UN-Water, 2018. The United Nations World Water Development Report 2018: *Nature-Based Solutions for Water*. Paris.
2. Jonathan Lautze and Herath Mathrithilake, 2012. *Water security: Old concepts, new package, what value?*. GWF Discussion Paper 1250, Global Water Forum, Canberra, Australia. Available online at: http://www.globalwaterforum.org/2012/11/20/water-security-oldconcepts-new-package-what-value/.
3. Maziotis, A., Molinos-Senante, M. and Sala-Garrido, R., 2016. *Assessing the impact of quality of service on the productivity of water industry: a Malmquist-Luenberger approach for England and Wales*.
4. Zeitoun, M., 2011. *The Global Web of National Water Security*. Global Policy **2(3): 286-296.**
5. Organization for Economic Co-operation and Development (OECD), *2013.Water Security for Better Lives*, OECD Studies on Water. OECD Publishing. http://dx.doi.org/10.1787/9789264202405-en.
6. World Bank, 2010. *The Cost to Developing Countries of Adapting to Climate Change*. Washington.
7. Schmeier S., 2013. Governing International Watercourses. *The Contribution of River Basin Organizations to the Effective Governance of Internationally Shared Rivers and Lakes*. London: Routledge.
8. United Nations University. Institute for Water, Environment & Health (UNU-INWEH), 2013. *Water Security & the Global Water Agenda*. Hamilton. Ontario. ISBN 978-92-808-6038-2.
9. Olli Varis , Marko Keskinen and Matti Kummu, 2017. Water & Development Research Group, Aalto University, Espoo, Finland, " *Four dimensions of water security with a case of the indirect role of water in global food security*", Water Security **1 (2017) 36–45**, Elsevier, 2017.
10. Grey D, Garrick D, Blackmore D, Kelmen J, Muller M, Sadoff C, 2013. *Water security in one blue planet: twenty-first century policy challenges for science*. Phil. Trans. R. Soc. A 371, 20120406. (doi:10.1098/rsta.2012.0406).
11. Jim Hall and Edoardo Borgomeo, 2013. *Risk-based principles for defining and managing water security*. Philos Trans A Math Phys Eng Sci. 2013 Nov 13; 371(2002): 2012040. DIO: 10.1098/rsta.2012.0407.
12. Mason N, Calow R., 2012. *Water security: from abstract concept to meaningful metrics: an initial overview of options*. London, UK: Overseas Development Institute.
13. Dustin Garrick and Jim W. Hall, 2014. *Water Security and Society: Risks, Metrics, and Pathways*. Annu. Rev. Environ. Resour. 2014. **39:611–39**, DOI: 10.1146/annurev-environ-013012-093817.

14. Water Sector Regulatory Council, 2018. *Bridge to Sustainability, Water and Wastewater Service Providers*, Performance Monitoring Report for the Year 2016, Ramallah, Palestine.
15. World Bank, 2018. *Toward Water Security for Palestinians: Wes Bank and Gaza Water Supply Sanitation, and Hygiene Poverty Diagnostic.*
16. Global Water Partnership Mediterranean (GWP-Med.), 2015. *Water Governance in Palestine: Sector reform to Include Private Sector Participation* (Report developed in the framework Governance and Financing for The Mediterranean Water Sector with support of SWEDEDN Med-Partnership in Partnership with OECD under the guidance of Palestine Water Authority).

Uses of Augmented reality for urban utilities management

Ahmed KADDIOUI [1,2], *Isam SHAHROUR*[1], *Ahmed El OIRRAK*[2]

[1]*Laboratoire de Génie Civil et géo-Environnement, Université de Lille, 5900 Lille, France*

[2]*Laboratoire d'ingénierie des systèmes informatiques, Université Cadi Ayyad, Marrakech, Maroc*

Abstract. Augmented reality (AR) is based on supplementing the view of the physical world with a computer-generated content such as text, images videos, sound, 3D objects etc. The proliferation of smart mobile devices and their high computational capabilities combined with the possibility of GIS 3D visualization and interaction have enabled new kinds of mobile applications. AR is considered as an efficient technology for the underground utilities management. Damages to underground utility lines are considered as one of the major urban problems because incidents caused by unsafe digging practices can compromise community safety and disconnect people from vital services. A prospective study was conducted to identify examples and characteristics of each type of augmented reality. This paper summarizes researches about the types of AR and discuss how this technology could help in underground utilities repairing and maintenance. We present a new mobile location based augmented reality system using smart device sensors and the Real Time Kinematic (RTK) satellite navigation technology of ORPHEON network for the location accuracy. This system enables GIS data visualization to help field workers in outdoor maintenance of underground infrastructures.

Keywords: Augmented Reality, underground infrastructures, buried networks, GIS 3D visualization.

1. Introduction

Cities face a multitude of challenges, because of the fast growth of the population. Underground utilities that provide drinking water, electric power, sewage, and telecommunications, among other services have been gradually added leading to complex subsurface problems. However, information about the localization and condition of underground infrastructures are generally not easily available, impacting their maintenance and the urban development. The efficient management of these infrastructures is still considered as an unresolved problem for both maintenance and installation companies [1]. Due to the lack of information about the localization of buried utilities, soil excavating for maintenance could lead to utilities damage with sometimes large consequences on urban services supply.

Generally, utilities companies use 2D maps for inspection of the underground assets, involving the tedious task of locating them correctly. Today, with the digital revolution, advanced systems for integrated management of buried networks could be used. These systems offer the possibility to conduct collaborative work for the design, construction and management of urban underground utilities. Tools for computer-aided design (CAD) are generally used in the design phase. However, these tools are not updated during the operating phase. We have also some difficulties related to crack detection to access the exact depth of some objects [2].

An integrated system for buried networks requires a client–server architecture. It should be able to extend the geographic information systems (GIS) features in terms of spatial databases and data analysis to data visualization objects and ubiquitous access. Augmented Reality (AR) constitutes a powerful tool for this requirement, because it provides both an integrated 3D view of the geospatial models and immediate feedback to users. It acts in an intuitive way for data capturing, data correction and surveying. Location awareness and visualization of invisible information are additional valuable features.

This paper presents the use of AR in the maintenance of buried urban utilities. First, it provides classification of the AR types, then presents some examples of AR uses in underground networks and finally proposes a novel AR system for the underground utilities management.

2. Classification of AR types

Augmented Reality is a real time, direct or indirect view of the physical "real-world" environment, which can be augmented by computer-generated sensory input such as sound, video, graphics or GPS and compass data [3]. The overlaid sensory information can be constructive (i.e. additive to the real environment), or destructive (i.e. masking of the real environment). AR is an interactive experience of the physical environment whose objects are enhanced by computer-generated perceptual information [4].

AR has five types that fall into two global categories: (i) Maker based AR which uses camera and visual cues and (ii) Marker-less based AR that use positional data such as mobile sensors data like GPS, gyroscope and compass.

2.1 Marker-based AR

Called also image recognition AR, Marker-based augmented reality anchors a computer-generated virtual contents to some forms of visual marker of the real life. A marker can be an image 2D or a Fudicial marker. A fudicial marker can be black and white images that consist of 2D image, barcodes or "quick response" (QR) codes [5] as shown in (Figure 1). Marker-based AR uses a camera with an AR software to detect AR markers as the location for virtual objects. The marker should be kept in front of the camera, once the marker is recognized by the camera frame, a virtual computer-generated content can be viewed on a screen and digital assets are placed into the scene at the location of the markers [6] (Figure 2).

a) Quick response code b) Barcode c) 2D image

Figure 1: Fudicial marker

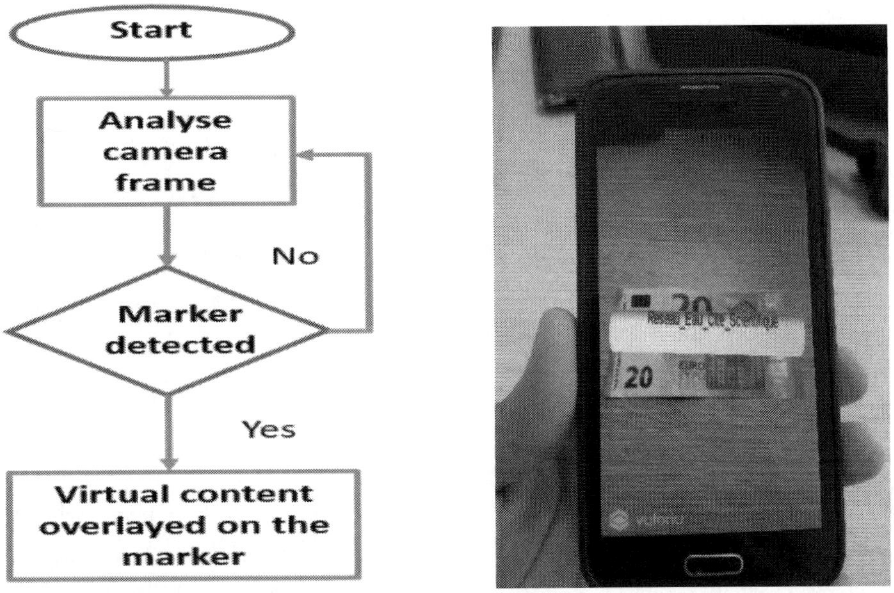

Figure 2: Marker-based augmented reality system

2.2 Marker-less based AR

The Marker-less AR technology doesn't need prior knowledge of a user's environment to overlay a 2D or 3D content into a scene and hold it to a fixed point in space. This technology uses data from camera systems, dedicated sensors and complex algorithms to accurately detect and map the physical environment — such as the location of points of interest. Drawing a map of the area, a Marker-less AR application enables placing virtual objects into a real context and have them in place without the need for any Fudicial marker [7] [8]. This technology includes different AR types such as Projection based AR, Location based AR, Superimposition based AR and Outlining based AR. Each type has specific characteristics (Table 1).

Category	Type	Examples	Characteristics
Marker based AR	Recognition based AR	- Augment: [9]	-Focuses on recognition of objects -Needs a camera to scan a barcode or QR code. -Add digital information to the object. - Once the object is recognized it can be replaced with another virtual object.
Marker-less based AR	Projection based AR	-Aerial Burton: [10]	-Functions using projection onto objects can be used to create deception about the position, orientation and depth of an object. -Might eliminate the use of special gear like head mounted displays
	Location based AR	- Yelp Monocle: [11] - Instagram: [12]	-Used for AR location browsers -Detecting the device location and orientation by reading its data sensors -Predicting where the user is looking -Adding related information on screen about specific objects that can be seen from the camera.
	Superimposition based AR	-Layar: [13]	Provides an 'alternate' view of the object in concern by: -Replacing the entire view with an augmented view of the object; -Replacing a portion of the object view with an augmented view.
	Outlining based AR	-110 Stories Kickstarter [14]	Solve human eye view limitations like: -looking at things for too long; -see well in low light conditions -Apps which perform outlining use special cameras.

Table 1. Summary of augmented reality categories and types.

3. Applications of Augmented Reality in underground networks

The lifecycle management of the underground utilities including design, construction and maintenance are generally accomplished using different techniques and data formats [15]. Design falls within the domain of the Building information modeling (BIM) and computer aided design (CAD). Generally, 2D modelling is used. Since the maintenance could alter the original design, CAD tools are not adequate for the comprehensive management of underground utilities. Consequently, integrated system that provides storage capacity, visualization, and analysis capabilities associated with the Geographic Information Systems (GIS) is required. Recently, BIM and 3D GIS were used for both design and spatial data during the planning and construction phases as well as the system lifecycle management [16].

Recent advances in sensors, embedded microsystems, internet and wireless communications have led to the evolution of a new generation of distributed computing platforms and resulted in innovative systems for data representation, interaction and acquisition. AR is an emerging technology that provides overlaying virtual content seamlessly on the real world and has a great potential in the field of underground utilities management. It allows to visualize the underground assets [17]. An advanced system based on AR was developed for the visualization of subsurface infrastructures. A tablet is used as a display, an inertial measurement unit is incorporated, as well as a GPS antenna and a camera. Sensors incorporated in the device "tablet" make it bulky. This AR system allows an interactive planning and surveying of geospatial data and use real time kinematic RTK for solving the location problem [18]. The use of physical targets allows device's orientation with an accuracy below centimeter. The difficulty of this technique is related to the necessity to figure out the target points, which is very complex in dense cities. [8] Proposed an AR system that allows creating, reading, updating, and deleting data from a geospatial database through a mobile interface which is connected to Internet by means of a client–server architecture. This system can be deployed easily using only underground infrastructures data in CAD format.

Chicago city has deployed a system for both city mapping and assessing of underground infrastructure [19]. Some utility companies use the Ground-penetrating radar (GPR) as a remote non-destructive sensing technique for both underground infrastructure imaging and assess buried pipe condition assessment [20, 21,22]. [23] Proposed a GPR system integration with AR mobile-based positioning system. This system uses fundamentals of GPR sensing and the Google Tango AR technology. It can perform a circular GPR scan over a distance of 50 m and assigning a geographic coordinate accurately to detected underground features.

4. Novel AR system for the underground utilities management

The proposed AR system presents new features. It combines different emerging technologies to develop an AR mobile device (Smartphone or Tablet) for buried networks management and using a client-server architecture (Figure 3).

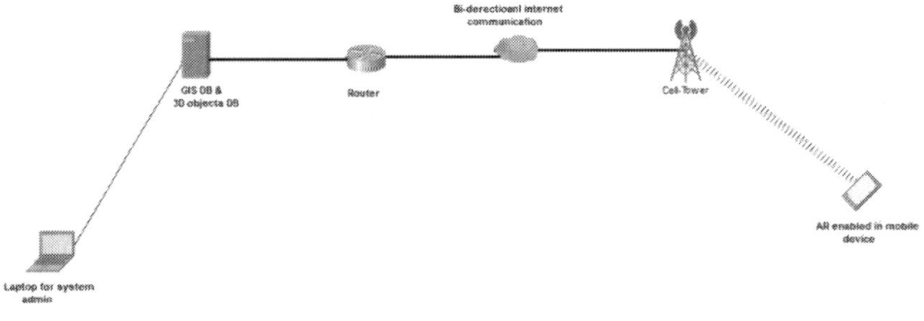

Figure 3: AR system Client–server architecture

This system integrates technologies in the domains of AR and 3D GIS. It aims at offering an AR Assistive System to help utility field workers to visualize the buried infrastructures on a Smartphone or a tablet Screen. This system provides a real time and high accuracy visualization of surrounding underground utilities (water, gas, sewerage, electricity, etc.). As this system is designed for maintenance of buried utilities, it should provide a localization accuracy. This is achieved by using the Real Time Kinematic (RTK) [24] satellite navigation of ORPHEON network [25]. This system provides an ergonomic easy to use interface (UI). The user could interact with the system and getting the results displayed on his device screen. The system allows bi-directional interactions between the GIS database, the 3D objects database and users. All changes on the GIS database would be automatically translated into 3D objects and synchronized with GIS databases and 3D objects database (figure 4). Users have informations precisely about the device position and orientation and the surrounding underground utilities.

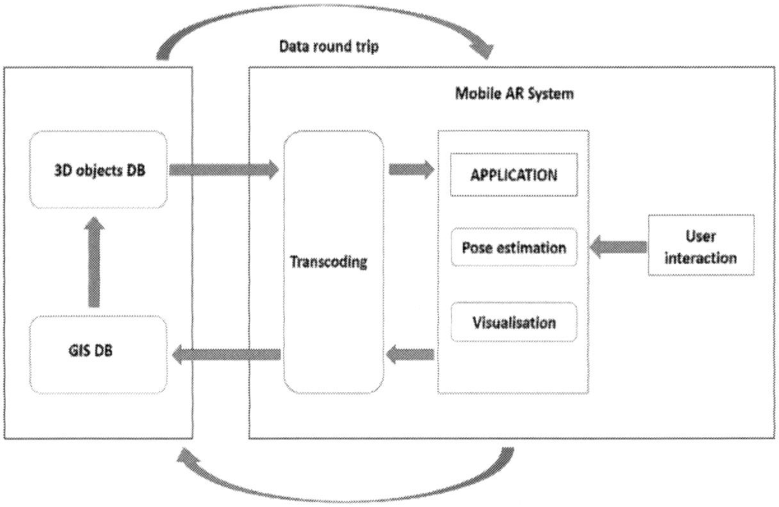

Figure 4: AR data round trip system

5. Conclusions

The management of underground utilities constitutes a major urban concern. Today, it constitutes an important component of the Smart City. Augmented Reality (AR) offers powerful capacities in the field for the visualization of underground utilities in their 3D environment as well as the associated real-time informations. This paper presented AR classifications and pointed out the critical issues related to the use of AR in utilities management. It also presented an innovative mobile AR system for utilities management using the smartphone sensors for the localization of the device and hidden utilities tracking and visualization. This system uses the Real Time Kinematic (RTK) satellite navigation technology of ORPHEON network, which is available in France with a localization accuracy within some centimeters. In addition, this system has the capacity to connect to a server and update utilities informations.

References

1. Gamayunova, O., and Gumerova, E., 2016, "Solutions to the Urban Problems by Using of Underground Space," Procedia Eng., 165, pp. 1637–1642.
2. Curiel-EsparzaCanto-Perello, J., 2013, "Selecting Utilities Placement Techniques in Urban Underground Engineering," Arch. Civ. Mech. Eng., 13(2), pp. 276–285.
3. P. Milgram and F. Kishino, "A Taxonomy of Mixed Reality Visual Displays," IEICE Trans. Information Systems, vol. E77- D, no. 12, 1994, pp. 1321-1329.
4. The Lengthy History of Augmented Reality". Huffington Post. 15 May 2016.
5. Craig, Alan B. « Augmented Reality Hardware ». In *Understanding Augmented Reality*, 69-124. Elsevier, 2013.
6. A. Katiyar, K. Kalra, et C. Garg, « Marker Based Augmented Reality », Advances in Computer Science and Information Technology, vol. 2, no 5, p. 5, 2015. pp. 441-445.
7. https://www.marxentlabs.com/what-is-markerless-augmented-reality-dead-reckoning/
8. K. Yang, T. Brown, et K. Sung, « AR Object Manipulation on Depth-Sensing Handheld Devices », Applied Sciences, vol. 9, no 13, p. 2597, June 2019.
9. https://play.google.com/store/apps/details?id=com.ar.augment&hl=fr
10. https://www.youtube.com/watch?v=1QXw3ylCYT0
11. (https://www.yelp.com)-
12. https://www.instagram.com
13. https://edshelf.com/tool/layar/
14. http://110stories.com
15. Sevket Demirci, Enes Yigit, Ismail H. Eskidemir, and Caner Ozdemir. 2012. Ground penetrating radar imaging of water leaks from buried pipes based on backprojection method. NDT and E International 47 (2012), 35–42.
16. Dryver Huston, Tian Xia, Yu Zhang, Taian Fan, Dan Orfeo, and Jonathan Razinger. 2017. Urban underground infrastructure mapping and assessment. April (2017), 101680M.
17. Amr Ahmed, Yu Zhang, Dylan Burns, Dryver Huston, and Tian Xia. 2016. Design of UWB antenna for air-coupled impulse ground-penetrating radar. IEEE Geoscience and Remote Sensing Letters 13, 1 (2016), 92–96.

18. G. Schall, S. Zollmann, et G. Reitmayr, « Smart Vidente: advances in mobile augmented reality for interactive visualization of underground infrastructure », Personal and Ubiquitous Computing, vol. 17, no 7, p. 1533-1549, oct. 2013.
19. https://www.cnet.com/news/chicago-maps-its-underground-maze-rahm-emanuel/
20. Yu Zhang, Anbu Selvam Venkatachalam, Dryver Huston, and Tian Xia. 2014. Advanced signal processing method for ground penetrating radar feature detection and enhancement. In Nondestructive Characterization for Composite Materials, Aerospace Engineering, Civil Infrastructure, and Homeland Security 2014, Vol. 9063.
21. Yu Zhang, Anbu S Venkatachalam, and Tian Xia. 2015. Ground-penetrating radar railroad ballast inspection with an unsupervised algorithm to boost the region of interest detection efficiency. Journal of Applied Remote Sensing 9, 1 (2015), 095058.
22. Tian Xia, Xianlei Xu, Anbu Vekatachalam, and Dryver Huston. 2012. Development of a High Speed UWB GPR for Rebar Detection. In 14th International Conference on Ground Penetrating Radar (GPR).
23. M. Pereira, D. Burns, D. Orfeo, R. Farrel, D. Hutson, et T. Xia, « New GPR System Integration with Augmented Reality Based Positioning », in Proceedings of the 2018 on Great Lakes Symposium on VLSI - GLSVLSI '18, Chicago, IL, USA, 2018, p. 341-346.
24. Real Time Kinematic (RTK) satellite navigation. http://en.wikipedia.org/wiki/Real_Time_Kinematic Last accessed 18 September 2016.
25. ORPHEON network http://reseau-orpheon.fr/le-reseau-orpheon/

Use of BIM and Smart Monitoring for buildings' Indoor Comfort Control

Rania Wehbe[1,*], *and Isam* Shahrour[2]

[1]Laboratoire de Génie Civil et géo-Environnement, Université de Lille, 5900 Lille, France

Abstract. Building information modeling (BIM) is the geometric way to present a life cycle construction project including geographic information. Recently, the Internet of Things (IoT) has been progressively used smart buildings in order to enhance living comfort, work productivity and entertainment. However, studies addressing the combination of these two technologies (BIM and IoT) focused on the automatic diffusion of data through sensors to BIM models [1]. Based on American College of Occupational and Environmental Medicine (ACOEM) a great portion of our time is spend inside buildings, in our offices, homes, schools, health care facilities, or in other private or public buildings. Hence the necessity to improve the basic human right to live in a healthy, safe and comfort environment is vital. This paper presents the use of BIM to support complex decisions concerning comfort conditions in buildings. This use is illustrated through a case study concerning a building of the AUST campus in Beirut.

1 Background

This paper deals monitoring of indoor comfort conditions in buildings. The significance of the subject lies on the impact of indoor conditions on public health and quality of life. Thanks to the development of smart monitoring based on IoT, both occupants and buildings managers could access to real-time data concerning the indoor comfort. Consequently, they have the possibility to take mitigation measurements in the case of deterioration of indoor conditions. Smart monitoring could help also to enhance the indoor security [2].

The real-time survey of indoor comfort conditions faces some difficulties such as real-time management and share of data, 3D visualization of comfort conditions and rapid decision making.

The BIM technology offers large opportunities for the lifecycle management of building [3]. However, it lacks the means to integrate data concerning the immediate environment into the 3D model, that can stock information and improve human comfort. It must be supported by a plug-in to attain the first stage in intelligent building. Kensek confirmed the efficiency of the scripting language Dynamo as a parametric design tool, and successfully used data from Arduino environmental sensors (humidity, CO2 and sunlight) in the BIM model, while changing 3D model parameters [4].

This paper presents the methodology followed to use of BIM in real-time control of indoor comfort conditions then it provides an illustration of these through an application to a building in the AUST campus in Beirut.

rania.a.wehbe@hotmail.com

2 Methodology

2.1 Use of BIM for survey comfort conditions

The methodology followed in this research is illustrated in fig.1. Autodesk Revit Software is used to connect the real time information to the BIM system. Data transfer to the BIM model is conducted through dynamo parametric design script. The key issue to manage and monitor the building is related to linking the Model on Dynamo. The latter is a visual programming tool that works with Revit. Dynamo extends the power of Revit by providing access to the Revit API (Application Programming Interface) in a more accessible manner. Rather than typing code, Dynamo generates a script by manipulating graphical elements called nodes.

BIM model will manage and control the indoor comfort parameters of the building. First of all, all existing sensors are linked to Arduino software that generate the data periodically in a CSV form. Using Dynamo Revit the software will be able to interact in real-time with the CSV file. This database will help the Revit by mean of Dynamo to locate the coordinates of the sensors, their readings and display colors depending on the comfort conditions.

Dynamo is capable of detecting the comfort parameters (temperature, humidity, brightness, noise) of each room in the model and compare the values with the standards. If the comfort parameters do not satisfy standards values, dynamo will send a message visualizing the 3D model containing the rooms affected by abnormal values. In addition, the user can to follow the sensors data.

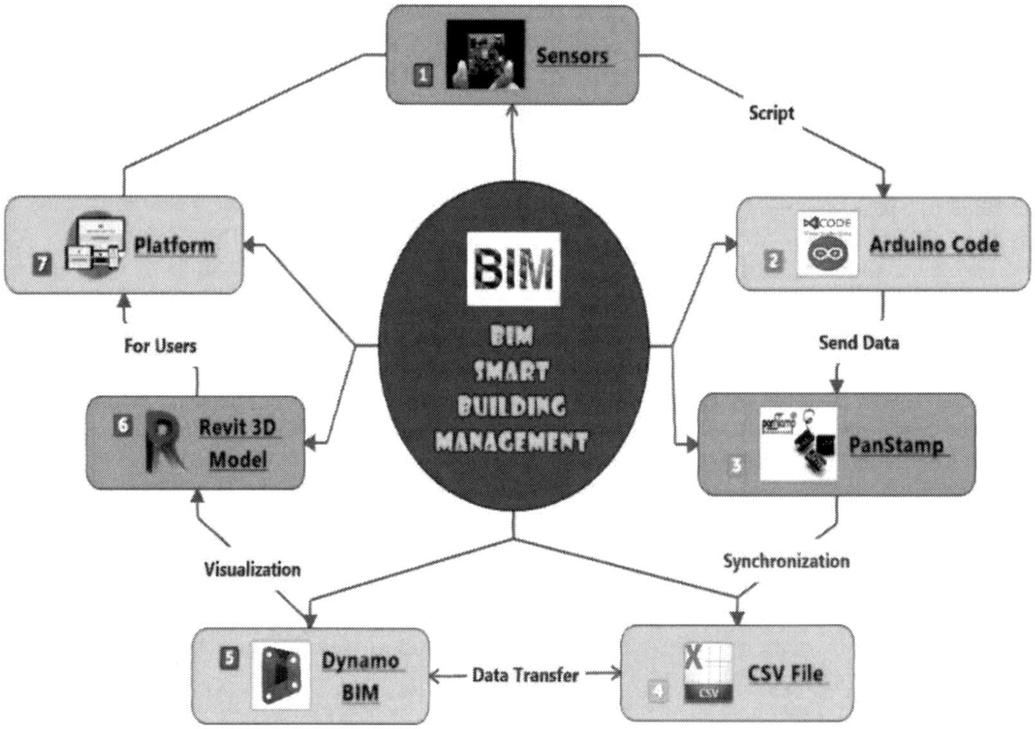

Fig. 1. BIM Smart Building Management Prototype.

2.2 Comfort parameters

The comfort parameters taken into consideration in this research concern the temperature, humidity, noise and brightness. The comfort indoor temperature for human beings is between 19 and 26 degrees Celsius.

On the other hand, even if the temperature is maintained and controlled between the ideal ranges, humidity in the air can affect the comfort in the buildings no matter what type of existing climate. Having the right level of moisture in the air ensure that the HVAC system in the building are working efficiently. The average humidity in summer should range between 30% to 45% while in winter, humidity should be below 40% [5]. Controlling humidity in buildings is very important, because the right humidity makes the air feel cooler in summer and warmer in winter, therefore it will increase the efficiency of heating and cooling system and reduce energy consumption.

Decisions concerning the indoor comfort are based on Predicted Mean Vote (PMV), knowing that the PMV is a common metric for measuring the comfort level of an indoor environment according to ISO 7730 [6]. PMV can be calculated based on factors including the clothing insulation and body's metabolic rate, besides environmental parameters (air temperature, humidity, average radiant temperature, and relative air velocity). PMV formula is written through Python in Dynamo to analyze PMV values on BIM Model. When PMV = 0, the indoor environment is in the best thermal comfort state, moreover the PMV should be between -0.5 and 0.5 based on code.

According to WHO large noises can affect largely human being comfort and productivity, because human beings to be prolific in calm environment. Based on Tocci Room Noise Criteria, indoor noises should not exceed 55 db [7].

The indoor light is essential and vary depending on the indoor human activity. However, the light level should be in the range of 500 to 1000 lux.

3 Application to building A of the AUST

The prototype was at first tested in building A in AUST university, faculty of engineering, for the temperature, humidity and brightness. Sensors record and transmit data at 5 minutes time-interval. According to standards, the temperature is acceptable, the room is slightly humide.

Table 1: Data collected from the sensors

Sensor Name	Time	Temperature (Celsius)	Humidity %	Brightness (Lux)
0A	14/06/2019 19:04:21	24.9	50.3	355
25	14/06/2019 19:04:25	24.5	53.3	266
0A	14/06/2019 19:09:22	24.8	49.9	290
25	14/06/2019 19:09:26	24.5	52.7	232
0A	14/06/2019 19:14:23	24.8	50.6	350
25	14/06/2019 19:14:26	24.4	53.5	252
0A	14/06/2019 19:19:24	24.7	51.1	267
25	14/06/2019 19:19:27	24.4	53.7	256

Table 1 shows an example of recorded data. The temperature is around 24°, while the humidity varies between 49.9 and 53.7%. The brightness varies between 252 and 355 LUX.

Fig. 2 shows the 3D BIM with several dynamo script based on different functions, sensor information, colors based on the PMV values, alert notification, charts representing recorded values.

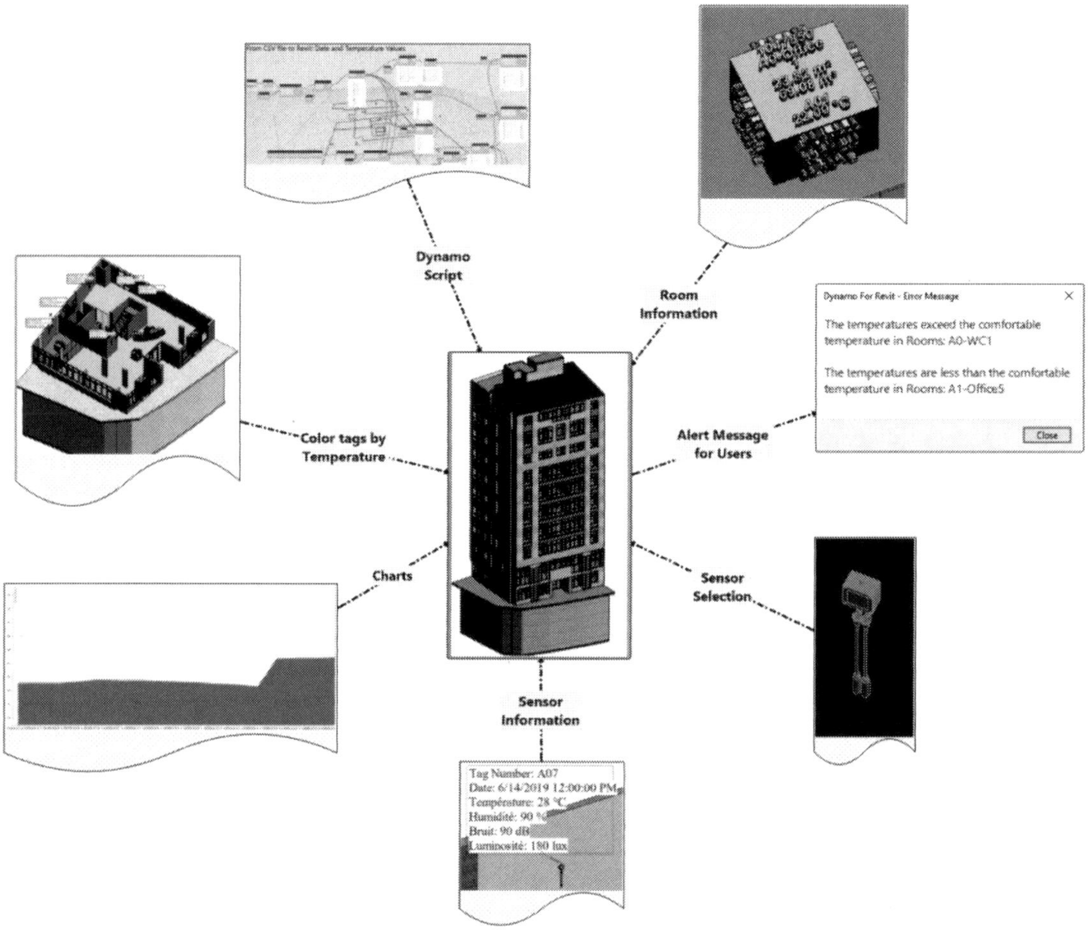

Fig. 2. Summary prototype function

Conclusion

This paper presented a system that combines BIM and smart sensors to control indoor comfort in buildings. This system allows a real-time survey of indoor comfort parameters as well as the 3D visualization of comfort parameters and the quality of these conditions with respect to buildings standards. The system was tested in Building A of the AUST campus in Beirut. The test showed good performances of this system in controlling the comfort conditions with respect to the quality of life and energy consumption.

Work is under progress to extend the capacity of this system to safety through monitoring of indoor hazards sources such as indoor air quality, electrical faults, fire, gas and water leak and space access.

References

1. K.-M. Chang, R.-J. Dzeng et Y.-J. Wu, Preprints **8**, 1086-1101 (2018)
2. R. Wehbe et I. Sahrour, MATEC Web Conf. **281**, 01013 (2019)
3. S. Azhar, M. Khalfan, T. Maqsood, AJCEB **12**, 15-28 (2012)
4. K. Kensek, Inf Constr, **66** (2014)
5. S. Unsdorfer, (2018). Available: https://www.centralhtg.com/blog/managing-home-humidity-for-maximum-comfort
6. ISO, Management **3**, 605-615 (2005)
7. C. Hansen, (2019). Available: https://www.archtoolbox.com/representation/architectural-concepts/architectural-acoustics-acceptable-room-levels.html

Progress on the exploitation of urban underground space (UUS) in China

Jie Liu[1], *Ling* Zeng[2], and *Hanbing* Bian[3,*]

[1]School of Traffic & Transportation Engineering, Changsha University of Science & Technology, Changsha 410114, Hunan, China

[2]School of Civil Engineering, Changsha University of Science & Technology, Changsha 410114, Hunan, China

[3]Laboratoire de Génie Civil et géo-Environnement, Université de Lille, 5900 Lille, France

Abstract. With the rapid urbanization in China, the disasters and the problems related with the urban underground space (UUS) exploitation increase. As an inevitable choice for sustainable urban development, how to increase the efficiency of the underground space use and augment the resilience of the UUS are the main challenges today in China. In this study, the strategic demands of the sustainable urban development on the underground space exploitation in Chinese cities have been analyzed and discussed. The associated problems and difficulties on the underground space exploitation were discussed. Moreover, the possible solutions were also proposed.

1 Introduction

The urban underground space (UUS) refers to the space below the ground available for human development and utilization [1, 2]. With the 30-year's rapid economic development, the urbanization has been sufficiently developed in China. In the same time, so many problems associated with the urbanization are also emerged. The most important is the shortage of the land resources and the green space in the urban area [3]. The exploitation of UUS resources is an inevitable choice for the sustainable urbanization.

* Corresponding author: hanbing. bian@univ-lille.fr

The China, as well as the human all over the world, has long history for underground exploitation. Not only for the space (habitation, reservoir, etc.), but also for the resources (water, energies, miners, etc.). However, these experiences are far away from todays demands. Comparing with the modern urbanization, the experiences for the exploitation of UUS in China is relative late. The first metro in China opened in 1971, while the first metro in London is 1863. Even though, the development of UUS is very rapid. The first line of metro in Shanghai is 1993. Today, the metro of Shanghai is longer and more complicated than that of London's. However, with the rapid development, so many problems related the exploitation of the UUS has been unveiled. These problems concern the Chinese legal system, the planning policies, the urban management, as well as the geo-technologies, the relative scientific researches, the industries equipment, etc. [4-6]. As consequence, so many disasters and accidents have been reported in recent years, and have a tendency to increase. In the developed country, as in European, during the rapid development, the same phenomenon has also been reported [7]. Today, with the climate changes and other challenges, the sustainable development should be taken into consideration. The new technologies, such as the smart city, the big data, the AI could be helpful for the UUS exploitation for stratify the citizens and the stakeholders [8, 9].

The main objective of the current work is to review the progress of the exploitation of UUS in China. Firstly, the strategic demands of underground space in Chinese cities are analyzed. Then, the problems and difficult are discussed. Finally, the possible solutions are proposed.

2 Strategic demand of UUS exploitation in China

In the past 30 years, the urbanization in China has recovered hundreds of year's history of the developed countries. It is also one of the most important engines for Chinese economy. The UUS exploitation is closely related to the urbanization [10, 11]. Today, China's underground space development has also many strategic demands:

2.1 Improvement of urban space resource utilization

Urbanization rate in China has increased from 20.16% in 1981 to 56.10% in 2015, and the national urban area increased from 7,000 km^2 to 49,000 km^2. Supposing an urbanization rate as 70%, 28,000 km^2 of land resource is needed [12]. Although, China has sufficient land resource for this development theoretically, most cities in China, either the quantities of population or the density are not sustainable for the development [13]. The lack of infrastructures is evident, from the transportation to the parking, to the green space, etc. The underground rail transit network and the underground expressway network can significantly alleviate the traffic pressure in the city and replace 5% to 8% of urban road. If the proportion of parking spaces and vehicles in China's increase to 1.3 times, 1000km^2 is needed in the cities. The underground construction of half of the newly built parking lot can save 600 km^2 of urban construction land. In addition, the underground and semi-underground of infrastructure (e.g. substations and sewage plants) can replace urban green space in large quantities.

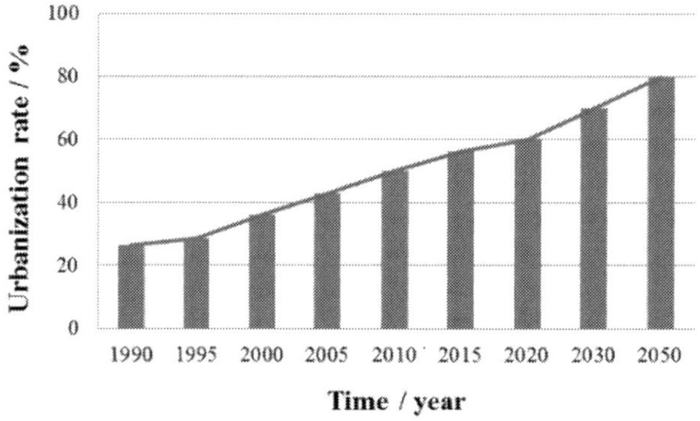

Fig. 1. Urbanization rate forecast in China

2.2 Urgent demands for solving "urban diseases"

Urbanization is an inevitable trend in the development of human society [14]. However, the increase in urbanization rate means that more people are crowding into cities, which brings great challenges to the development of cities and is the main cause of "urban diseases". Developing UUS will help to solve the problems that Chinese cities are currently facing.

(a) Traffic jams and parking. Traffic jams is the common problems of Chinese cities, from large cities, as Beijing and Shanghai, to the small cities, the jams are frequently. As the density of population is relatively high in the urban area, the jam is inevitable. The underground transportation system, especially the public transportation system (metro, tramway, the underground bus network, etc) could be good solution. The underground parking is an alternative choice, not only for increasing the parking place, but also for the urban green area.

(b) Air pollution and urban environment. Automobile is currently considered the main source of air pollution and the greenhouse gas in major cities in China. It has great potential and significance in alleviating the air pollution problems and improving urban environment in China by developing underground space public transportation system and underground network.

(c) Improving infrastructure. Due to the low level of urbanization and the rapid expansion of cities, the quality of urban infrastructure needs to be improved in China. With the ambition projects of technical galleries (underground utility tunnel), all the systems (including, electricity, gas, water, communication, light, heat, etc.) has been integrated in the same gallery. The underground utility tunnel, not only facility the infrastructures' maintenance, but also increase their safety.

2.3 Promoting urban economic growth

In the past 30 years, the urbanization accelerated the economic growth. In the next years, it is the UUS, the new kinetic energy to stimulate economic growth, promotes industrial development and creates new jobs. In 2020, the urban rail transit system will double by that in 2015, and the employs will increase by 18%. By 2035, the total urban rail transit system in China is expected to reach or exceed 15,000 km. These will undoubtedly encourage the sector of construction materials (e.g. steel and cement), machinery and equipment, and construction and management industries [15]. It is estimated that the output value of underground space utilization will reach 5~6 trillion yuan in 2030, which will form a new economic growth point in China's new era and help to improve urban employment and income levels.

3 Issues of UUS exploitation in China

The use of UUS before the reform and opening up of China is mainly based on the construction of civil defense projects (or military projects), with limited scale and single type. The use of UUS after the reform has entered an accelerated development stage, especially the rapid development of subway projects [16]. Since 2000, China has become the fastest growing country in the exploitation of underground space over the world. However, compared with the trend of exploitation of underground space in developed countries, there are still some gaps in the planning, conception, design and exploitation of the UUS.

3.1 UUS development function lags behind and lacks development depth

Nowadays, the exploitation of UUS in China is very fast, but the overall progress is still incomparable with the rate of urbanization [17, 18]. In addition, the underground space utilization is relatively simple. Developmental stages of UUS in China were shown in Table 1. Nearly 60%~80% of them are mainly used for underground parking. The underground space is arranged in a scattered point arrangement, and is not connected to each other. Moreover, most of UUS are within a depth of 30m, which will cause some constrains for further development. The main reason is that the master plan in most cities in China is far from meeting the needs of underground space planning and development, and the basic geological work supporting the development and management of underground space is still relatively weak. The current regional survey work is not comprehensive enough to grasp the geological characteristics, structure and geological parameters of deep underground space, which still has a big gap between the requirements for the transparent display of geological bodies. In addition, the sophisticated detection technology of deep underground space resources needs to be further studied, and it is still unable to fully meet the needs of supporting the development of underground space planning.

Table 1. Developmental stages of UUS in China

Developmental stage	Function	Characteristic	Depth /m	Representative city
Initialization	Underground parking, civil defense	Single unit construction, single function and small scale	<10	Prefecture-level cities
Large-scale	Rail traffic	Linear development along rail traffic	15~20	Qingdao, Xiamen
Initial network	Rail traffic nodes and complexes	The subway line begins to cross the key utilization nodes and underground complexes	30	Shenzhen, Tianjin
Scale network	Various public facilities	The subway network system is perfect, including commercial, transportation and municipal.	50	Shanghai, Beijing
Underground eco-city	Various underground facilities are integrated	A fully functional and ecological ecosystem	200	—

3.2 UUS information isolation

Compared with urban ground information management, the information of UUS involves planning management, land management, construction management, civil defense management, operation and maintenance management, environment and fire management, and so on. At present, the information management of urban municipal pipelines, integrated pipe corridors, underground rail transit, underground buildings, etc. is still relatively backward [19]. The integration of urban geological work and urban construction management is insufficient. Furthermore, there are information barriers between the technical and departmental departments, and the management system has not opened up the connection channels. Therefore, it is difficult to effectively provide scientific and practical urban geological basic data and achievements for urban construction.

The management of underground space in typical cities in China involves more than 20 units and departments, which presents a more dispersed and complex pattern and lack a clear coordination and coordination department [20]. Each management department assumes its own management functions. What's more, different departments have different management requirements, which may lead to complicated approval procedures, chaotic management, and unclear interest relationships. Therefore, the policies need to be improved to encourage the UUS exploitations.

3.3 The underground space risk assessment and resilience

The construction of underground projects will inevitably affect or even destroy the surrounding structures, geotechnical bodies and groundwater systems. The development of underground space faces different risks. In China, a complete risk assessment mechanism has not yet formed. The UUS exploitation is likely to cause a series of environmental and geological problems, such as ground deformation, land subsidence, groundwater pollution, etc. The occurrence of such problems is mainly due to the fact that the management system, the legislative issues, the standard, the norms for the exploitation of underground space in China are not yet sound enough to provide safe and effective guarantee for the exploitation of UUS. Among them, underground space planning, operation management and control planning, resource utilization policy guidance, three-dimensional development of land use planning system, underground space zoning development and guidance strategy, underground space land use planning indicators and planning implementation guarantees need to be further strengthened and improved.

4 Development trends of UUS in China

Nowadays, the urbanization in China will continue with the economic development. The exploitation of UUS is evident. In the same time, with the confrontation of many challenges and difficulties, the possible solutions adapting current situations should be proposed.

4.1 Layered and networked planning layout

With the rapid and separate development in the UUS in China, the most urgent problem for the further development in UUS is how to integrate the newly developed underground spaces with the existing underground buildings, generally separate with each other. Therefore, the long period guideline for a given city is particularly important. At the level of urban master planning, each city will combine the local geological characteristics and the direction of urban extension according to the future economic conditions and development needs. The UUS exploitation planning extends in not only the horizontal direction, but also the vertical direction. The UUS is discrete in different functional area with in the depth. Schematic illustration of layered development of underground space was presented in

Figure 2 [14]. In fact, the layout of the future underground space is more possible as network. For the developed countries, they encounter also the same problems from last century. The experiences and the lessons could be helpful for the Chinese UUS exploitation. For example, the metro, the utility tunnel, the underground commercial center, etc.

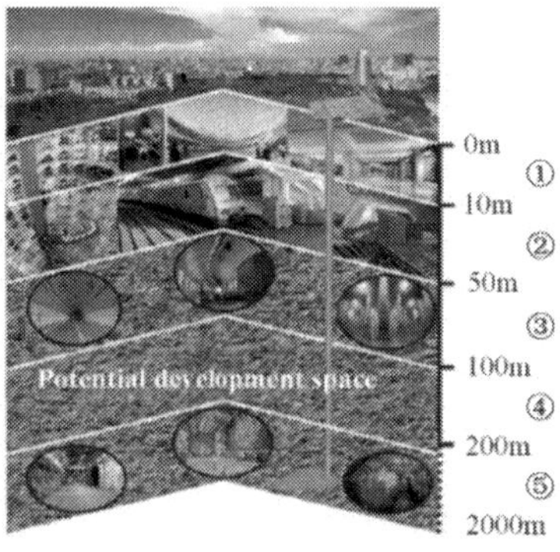

Fig. 2. Schematic illustration of layered development of underground space

4.2 Humanization

More considerations on the environmental characteristics of underground space and on the impact of underground space environment to human physiological and psychological levels [21] should be taken. As surrounding by the geo-formation, the UUS is considered as a closed space with limited accesses. The advantage is that it has good thermal isolation and consumes relative small quantity of energy for heating. The disadvantages are that the quality of the air and the humidity should be controlled for assuring the comfort and the safety of the user. The ventilation system is necessary, not only for the potential fire, but also for the daily operation and maintenance. With the increase in quality of life, besides of the safety in the UUS, the conformable is not the critical problem in the exploitation of UUS.

4.3 Digital and intelligent management

Large-scale construction and operation of UUS will generate massive data (e.g., planning, construction, operation, maintenance). At present, the digital and intelligent management in Chinese UUS systems is mainly based on GIS/BIM technology. However, GIS/BIM technology has some problems when it is applied in underground space, such as lack of geological model, analysis function, and data collection, difficulty in information transmission, collaborative work, and file compatibility. Strengthening the development of digital and intelligent management is crucial to the exploitation of UUS.

The planning, design, construction, operation, maintenance, disaster prevention and safety management of UUS are inseparable from comprehensive understanding and accurate of basic data of UUS. The use of digital and intelligent methods to manage UUS data is an inevitable development trend. It is a multidisciplinary and interdisciplinary problem, and beyond the classic engineering problem. Especially for the construction and decision-making of UUS in nowadays, as well as the urban disaster prevention and emergency response in the underground space, which has a tendency to be more and more complicates [22, 23]. The digital technologies and the conception of smart give the possible for the further development of the UUS. With the modern tools, such as the BIM, GIS, the AI, the digital and intelligent management system will be implemented in phases and improved in practice. The resilience of UUS also will be augmented.

5 Conclusions

The paper has introduced an overview of progress on the exploitation of UUS in China, and the following conclusions can be drawn:

(1) Exploitation of underground space in China has many strategic. The exploitation of UUS can effectively improve the utilization of urban space resources, which is an urgent need to meet the challenges in the urbanization, and favorite for the durable urban economic growth.

(2) China has become the fastest growing country in exploitation of underground space in the world. However, there are still many deficiencies and problems in exploitation of UUS. Many efforts should be paid for planification, the information isolation and the risk assessment and resilience.

(3) The lessons and experiences learned from developed countries could be helpful for the China in the UUS exploitation. The future development of UUS exploitation in China should consider layered and networked planning, humanization, etc. The digital and intelligent management is inevitable in the further UUS exploitation.

References

1. D.V.L. Hunt, L.O. Makana, Jefferson I., C.D.F. Rogers, Tunn. Undergr. Space Technol. **55**, 8–20 (2016)
2. R.L. Sterling, H. Admiraal, N. Bobylev, H. Parker, J.P. Godard, I. Vähäaho, T. Hanamura, Proc. Inst. Civ. Eng. Urban Design Plan. **165** (4), 241–254 (2012)
3. X.Z. Li, C. Li, A. Parriaux, W.B. Wu, H.Q. Li, L.P. Sun, C. Liu, Tunn. Undergr. Space Technol. **55**, 59–66 (2016)
4. Q.H. Qing, Tunn. Undergr. Space Technol. **55**, 280–289 (2016)
5. Z.L. Chen, J.Y. Chen, H. Liu, Z.F. Zhang, Tunn. Undergr. Space Technol. **71**, 253–

270 (2018)
6. J. Liu, Y. Li, G.J. Zhang, Y.N. Liu, Constr. Build. Mater. **227**, 116654 (2019)
7. P. Duffaut, Undergr. Space **5**(2), 86–91 (1980)
8. L. He, Y. Song, S. Dai, K. Durbak, Tunn. Undergr. Space Technol. **32**, 168–179 (2012)
9. W. Broere, Tunn. Undergr. Space Technol. **55**, 245–248 (2016)
10. N. Bobylev, Land Use Pol. **26** (4), 1128–1137 (2009)
11. C.D.F. Rogers, Eng. Geol. Spec. Publ. **22** (1), 177–188 (2009)
12. K. Malik, *Report of China human development*, (UNDP in Asia and the Pacific, Beijing, 2013)
13. National Bureau of Statistics (NBS), *Yearbook of Chinese cities*, (China Statistics Press, Beijing 1985-2015) (in Chinese)
14. X.Z. Li, R. Wang, Q. Gu, D.K. Zhou, Y.X. Xin, Earth Sci. Front. **26**(3), 32–38 (2019) (in Chinese)
15. C.S. Wang, C.H. Zhou, J.B. Peng, J. Fan, H.H. Zhu, X.Z. Li, G.H. Cheng, C.S. Dai, N.X. Xu, Earth Sci. Front. **26**(3), 1–8 (2019) (in Chinese)
16. B.C. Liu, Chin. J. Rock Mech. Eng. **18**(1), 109–111 (1999) (in Chinese)
17. J.B.M. Admiraal, Tunn. Undergr. Space Technol. **21** (3–4), 464–465 (2006)
18. P. Belanger, Tunn. Undergr. Space Technol. **22** (3), 272–292 (2007)
19. Z.H. Lv, X.M. Li, B.Y. Zhang, W.X. Wang, Y.Y. Zhu, J.X. Hu, S.Z. Feng, IEEE access **4**, 407–415 (2016)
20. G.H. Chen, R. Wang, M.H. Zhao, J.W. Su, Y. Yang, X.B. Zhang, Earth Sci. Front. **26**(3), 39–47 (2019) (in Chinese)
21. N. Ivanova, O. Ganzha, Proce. eng. **165**, 1006–1015 (2016)
22. M. Casals, M. Gangolells, N. Forcada, M. Macarulla, A. Giretti, M. Vaccarini, Appl. Energy **166**, 150–164 (2016)
23. P.P. Nelson, Tunn. Undergr. Space Technol. **55**, 32–39 (2016)

Optimization of sensors locations in internal stability analysis of Geosynthetic-reinforced earth retaining walls

Hicham Alhajj Chehade[1,2], *Marwan* Sadek[2], *Daniel* Dias[3,4], *Fadi* Hage Chehade[2], *Orianne* Jenck[1]

1 Laboratoire Sols, Solides, Structures, Risques, Université Grenoble-Alpes, 38000 Grenoble, France
2 Laboratoire Centre de Modélisation, Université Libanaise, Beyrouth, Liban
3 Université de technologie de Hefei, Hefei, China
4 Antea Group, Antony, France

Abstract. This paper concerns the optimization of sensors locations used to monitor the geosynthetic reinforcement internal forces of a reinforced earth retaining walls. The internal stability analysis of these structures is addressed through the kinematic theorem of limit analysis combined with the discretization technique to generate the failure surface. Knowing that the majority of damages of these structures are caused by the water presence in the reinforced zone, different water table levels are considered in the study and their effects on the critical failure surface location are analyzed.

1. Geotechnical structures risks

1.1 Geosynthetic reinforced earth retaining walls

Geosynthetic reinforced earth retaining walls is a construction technique based on the association of compacted backfill soil and geosynthetic reinforcement (straps, geogrids or geotextiles) connected to a facing (Figure 1). The structure behaves like a coherent flexible block, capable of supporting important loads thanks to the soil-reinforcement interaction.

The reinforced earth technique constitutes one of the most important development in civil engineering in the last five decades. This success is explained by the advantages offered by this technique compared to their conventional counterparts concerning money, time and space saving. These advantages have led to a wide use of this technique in many fields of Civil Engineering. It is used in the transportation construction in the roads, highways, bridges and railway structures, as well as for industrial and protective structures, dams, mining structures, in addition to its use for commercial and public structures. The multiple fields where reinforced earth retaining walls are used, show the importance of these structures in any urban development especially in areas with challenging design conditions.

1.2 Vulnerability of the reinforced earth walls

The reinforced earth walls suffer from vulnerability to natural hazards such as earthquakes and rainfalls. Safety constitutes the great challenge of this space. The vital role of geosynthetic reinforced earth walls, explained by their multiple applications, shows the importance of their management and monitoring. The failures of these structures could lead to huge human and economic losses. Many challenging issues have to be addressed when dealing with these structures such as very large structural loads, water presence, crack-formation in the backfill and seismic loading.

1.3 Rainfall hazard

In most important design codes (e.g. FHWA-NHI-11-024, NF P 94-270), a cohesionless soil is recommended and considered with an appropriate drainage system. These recommendations aim to ensure a fast removal of excess water from the soil. Then, the stability assessment of reinforced earth walls is conducted assuming the absence of pore water pressure in the backfill. However, most failures encountered in the literature were caused by the water presence in the backfill when a cohesive soil is used. Koerner and Koerner [1] presented in their paper the cases of 320 reinforced earth walls failures. Most of these failures are caused by internal or external water (201 cases or 63%), and in 232 cases (73%) a cohesive soil (Silt or clay) is used as backfill in the reinforced zone. This statistic study proved that the design codes specifications and recommendations are not respected in a lot of times. In fact, the circumstances sometimes lead to the use of poorly cohesive soils in the reinforced zone when appropriate granular soils are not available or expensive. Any malfunction of the drain system could then cause the development of the pore water pressure in the backfill and therefore, reduced the soil shear strength since the cohesive soil permeability is very low.

The water presence in the reinforced backfill is an important concern. Its effect depends from the water table level. It represents a serious risk on the internal stability of geosynthetic-reinforced earth retaining walls. The use of sensors on the geosynthetic reinforcement can help monitoring and consequently preventing risks that can occur.

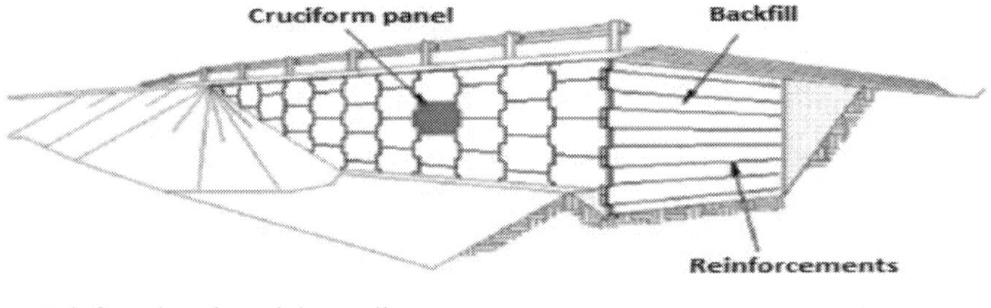

Figure 1: Reinforced earth retaining wall

2. Sensor optimization

2.1 Internal forces monitoring sensors

The installation of sensors on the geosynthetic reinforcement helps to monitor and measure the geosynthetic reinforcement internal forces. This leads to observe the evolution of the structure response and behavior, and prevent damages at early stage.

The internal stability of reinforced earth retaining walls is commonly investigated by using the limit equilibrium method. It is the most popular method due to its simplicity. However, it does not consider the stress-strain soil relations and it requires some assumptions. Limit analysis approach is an alternative methodology that considers the soil stress strain relation and permits to obtain rigorous upper or lower bound to the real solution through its two theorems. The upper bound theorem known also as the kinematic approach of limit analysis, is now widely used to investigate the internal stability of reinforced earth retaining walls. The form of failure surface is assumed in this kind of analysis. The failure surface is considered log-spiral in this study, based on the results and observations of previous works which indicate that it is the most critical failure mechanism.

In general, the reinforcement internal force varies along the geosynthetic element. The internal forces distribution is generally characterized by a maximum value at or near the intersection of the reinforcement with the critical failure surface. The reinforcement internal force or strain decreased gradually to both sides. It decreases to zero at the rear end (free end) and to the connection strength at the wall reinforcement connection.

To optimize the sensor locations in the reinforced zone in order to decrease their number, it is important to locate the regions where the critical surface could propagate. In unsaturated soil with variable water table level, the variation of the water table level may affect the anticipated failure surface location.

2.2 Internal stability of reinforced earth retaining wall

The kinematic theorem of limit analysis is based on the work rate balance between the external forces and the internal energy dissipation. It is used to find a lower bound to the reinforcement strength required to ensure the structure stability. The analysis here concerns reinforced earth walls where geosynthetic reinforcements are horizontal and uniformly distributed, have uniform length and their resistance to shear, bending and compression is ignored. Applicability of the theorem requires that the soil is described by an ideally rigid perfectly plastic behavior according to normality flow rule. The point-to-point method is used to generate the log-spiral failure mechanism. This technique has advantages over the conventional kinematic approach of limit analysis that it can be applied for non-homogeneous backfill. In addition, the pseudo-dynamic approach can be used to represent the seismic loading. However, the internal stability assessment is conducted here under static loading.

The generation of the failure mechanism is shown in Figure 2 where the block $ABDC$ rotates around the center of rotation O with an angular velocity Ω. The failure surface meets the velocity vector with an angle equal to the soil friction angle φ'. The angle $\delta\theta$ defines the accuracy of the failure mechanism. The positive parameter ξ defines the depth of the water table. When this parameter is equal to zero, the backfill is saturated. A value between 0 and 1 indicates that the backfill is divided in two zones, saturated and unsaturated. When this parameter exceeds 1, the backfill is unsaturated. The steps of the generation process and the work rate calculations are described in [2] who investigate the seismic internal stability of geosynthetic-reinforced earth wall with the pseudo-static method. In addition, the same authors include the crack formation as a part of the failure mechanism in their analysis in another study [3] where also, the pseudo-dynamic method is implemented to consider the seismic loading. However, in both works, the presence of water is not considered in the analysis. Since this paper considers only the static loading, the external forces contain only the soil weight and the pore-water pressure work rates. The second occurs only in the saturated zone. On the other hand, the internal energy dissipation occurs along the velocity discontinuity and along the reinforcement by tensile or pullout failures. For unsaturated zone, the soil suction tends to increase the soil cohesion. Therefore, the effective cohesion c' is used in the calculation of internal energy dissipation in the saturated zone (along BD). The total cohesion, resulting by the summation of the effective cohesion to the apparent cohesion due to matric suction, is used in the unsaturated zone (along DC).

The effect of water presence in the saturated zone on the external work rate calculation and the expression of the total cohesion in the unsaturated zone are presented hereafter.

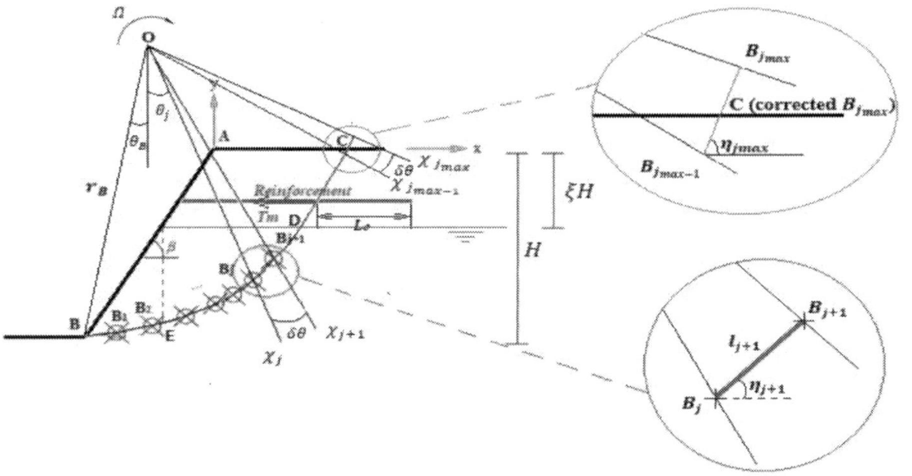

Figure 2: Failure mechanism generation process by the discretization method

In the saturated zone, the pore-pressure work rate considered as external force, is applied to the soil skeleton and the boundary. Since rigid block assumption is considered, the pore-pressure effect is limited at the boundary surface and it is calculated as follows:

$$\dot{W}_u = - \int_{S_{BD}} u.n_i.v_i dS \tag{1}$$

where v_i the velocity vector in the collapse block, S the boundary surface of the failure block respectively, n_i the outward unit vector normal to the surface S. The pore water pressure is considered by using the pore-water pressure coefficient r_u introduced in [4]. The pore-water pressure u at a depth h below the water-table level, is assumed to be equal to $r_u\gamma h$ where γ is the soil unit weight. The work rate of the pore water pressure along the discontinuity surface is given in the discretization method as the summation of the elementary work rate as follows:

$$\begin{cases} \dot{W}_u = \sum_j r_u.\gamma.(x_j.\tan\beta - y_j).l_j.\Omega.L_j.\sin\varphi' & \text{along the boundary } BE \\ \dot{W}_u = \sum_j r_u.\gamma.(-y_j - \xi H).l_j.\Omega.L_j.\sin\varphi' & \text{along the boundary } ED \end{cases} \tag{2}$$

where l_j is the length of $[B_{j-1}B_j]$, L_j the length of $[OB_j]$ an (x_j, y_j) the coordinates of B_j.

For unsaturated soil, the existence of the soil suction tends to increase the soil cohesion and consequently the shear strength. Several equations have been proposed to calculate the increase in the unsaturated shear strength τ_f with the matric suction $(u_a - u_w)$ with u_a and u_w being respectively, the pore air pressure and the pore water pressure. As proposed in [5], a linear equation for the unsaturated soil shear strength is considered. The linear increase of the shear strength was designated using the angle φ^b that represents the rate of increase in shear strength relative to matric suction. This parameter had the character of a friction angle. Then, the unsaturated soil shear strength can be expressed as an extended Mohr–Coulomb shear strength equation :

$$\tau_f = c' + (\sigma_n - u_a)\tan\varphi' + (u_a - u_w)\tan\varphi^b \tag{3}$$

where $(\sigma_n - u_a)$ is the net normal stress on the failure plane.

Equating the external work rates and the internal energy dissipation provides the objective function to be maximized in order to calculate the reinforcement strength required to ensure the wall stability. The geometrical parameters r_B and θ_B are considered in the optimization process, where r_B is the length of OB, θ_B is the angle between the y axis direction and the line OB as shown in Figure 2.

2.3 Results

In order to investigate the effect of water-table level and variations of hydrological conditions of the backfill soil on the required reinforcement strength and on the location of the critical failure surface and consequently the locations of maximum internal forces, a numerical example is presented here. The reinforced earth retaining wall is described by the following parameters: H=6 m, β=75°, γ=19kN/m^3, φ'=25°, c'=8 kPa, $\delta\theta$=0.01^o, L_r=0.75H, n_r=10, where H is the wall height, β is the wall inclination, L_r is the reinforcement length, n_r is the reinforcement number uniformly distributed.

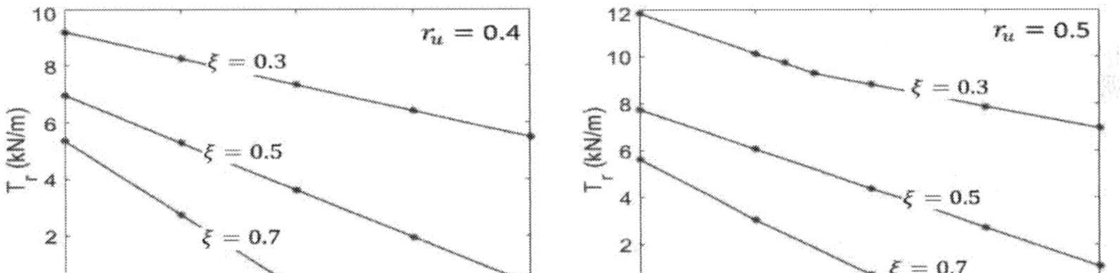

Figure 3: Required reinforcement strength for different value of soil suction

The results are illustrated in Figure 3. It is clear that the required reinforcement strength T_r decreases linearly with the soil suction in the unsaturated zone. It increases when the water table depth decreases and when the pore-water pressure coefficient r_u increases. The variation of the required reinforcement strength with the matric suction is linear except the case when the water table is at 1.8 m below the ground surface ($\xi = 0.3$) and the pore-water pressure coefficient $r_u = 0.5$. The reason of non-linearity is that under these conditions and for a matric suction smaller than 30 kPa, the geosynthetic reinforcements in the upper layers fail in pullout contrary to the other cases where all the reinforcement layers fail in tensile rupture. The results show the importance of accurate measurement and prediction of the soil suction and the importance of considering the water table level variation that can occurs due to the rainfalls.

During the structure monitoring, the sensors used to measure the internal reinforcement force must be implemented near the critical surface and when the reinforcement force measured approaches the anticipated required reinforcement strength, the monitor engineer must react because the limit state is reached and failure will occur. However, the variation of the water table level may change the location of the critical failure surface. If all the reinforcement layers fail in rupture (Figure 4a), the location of the critical failure surface remains almost the same. However, if some reinforcement layers fail in pullout (Figure 4b), the critical failure surface tends to drift away from the wall facing and the required reinforcements strength to maintain the structure stability is greater than the case when all reinforcement layers fail in tension. Thus, the critical failure surface location is influenced by the water table level and the soil matric suction in the unsaturated zone. For an important soil matric suction, the critical failure surface remains almost in the same location with the variation of water table level. However, it changes for small matric suction values. For an accurate monitoring and safety concerns, the sensors locations must consider the possible variation in the water table level and consequently cover a wider region predetermined. An alternative efficient method for monitoring the structure is to use simultaneously sensors to detect the location of groundwater. Once it is determined, the region of interest where the internal reinforcement strength sensors need to get attention is specified.

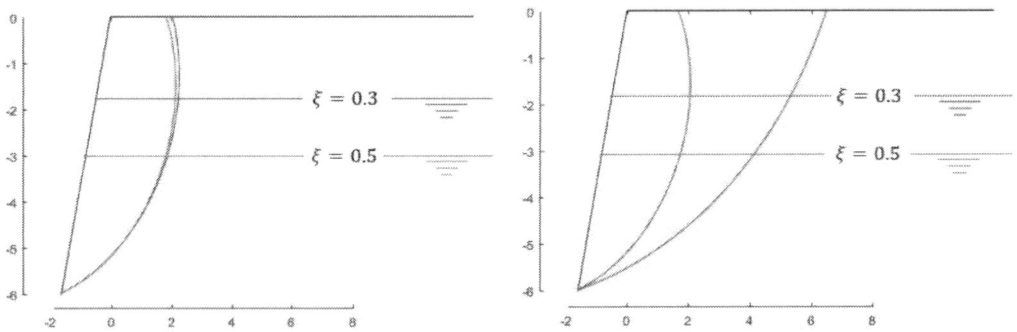

Figure 4: Critical failure surfaces for r_u=0.5: (a) $u_a - u_w = 60\ kPa$; (b) $u_a - u_w = 20\ kPa$

3. Conclusions

This work adopted the upper bound theorem of limit analysis combined with the discretization method for the determination of the reinforcement strength required to ensure the stability of geosynthetic reinforced earth retaining walls. The effect of water with different water-table levels are considered.

The results show the effect of the water table level and the matric suction in the unsaturated zone on the required reinforcement strength. In addition, the variation of the water table has shown a great influence on the location of the critical failure surface where the maximum internal reinforcement strength will occur. This happens when the upper reinforcement layers fail in pullout due the water table approach from the ground surface. The monitoring sensors should be implemented over a wide zone on the reinforcement layers to consider the possible variation of the groundwater level and the region of interest where the internal reinforcement strength sensors need to get attention is specified based on the location of the groundwater.

References

1. R. M. Koerner, G. R. Koerner,. Geotext and Geomembranes, **46**, 9 (2018).
2. H. Alhajj Chehade, D. Dias, M. Sadek, O. Jenck, F. Hage Chehade, Acta Geotech, (2019). *(Under review)*.
3. H. Alhajj Chehade, D. Dias, M. Sadek, O. Jenck, F. Hage Chehade, Geotext Geomembranes, **47**, 12 (2019).
4. A. W. Bishop, N. R. Morgenstern, Géotechnique, **10**, 25 (1960).
5. D. G. Fredlund, N. R. Morgenstern, R. A. Widger, Can. Geotech. J., **15**, 9 (1978).

Investigation of rock bolting for tunnels based on an efficient CATIA-ABAQUS model

Hui Lee[1,2], *Wei-zhong* Chen [1,3,*], *Xian-jun* Tan [1], *Qiao-chu* Yang [4], and *Xiao-gang* Wang [1,2]

[1]State Key Laboratory of Geomechanics and Geotechnical Engineering, Institute of Rock and Soil Mechanics, Chinese Academy of Sciences, Wuhan 430071, China
[2]University of Chinese Academy of Sciences, Beijing 100049, China
[3]Geotechnical and Structural Engineering Research Center, Shandong University, Jinan, Shandong 250061, China
[4]BIM Research Center, Sichuan Highway Planning, Survey, Design and Research Institute Ltd, Chengdu, Sichuan 610000, China

Abstract. As a widely used reinforcement method in tunnel engineering, the accuracy of models in numerical analysis has a vital influence on the reliability of the analysis results. To solve the problem of hard modeling for complicated models, this paper proposes to build models of different characteristics automatically by the means of Enterprise Knowledge Language (EKL) of CATIA, combining knowledge engineering template with parameterization. Furthermore, the stability of surrounding rocks and the distribution features of plastic areas were studied in different cases by CATIA-ABAQUS model. Besides, the effects and contributions of several parameters, including the rock-bolt diameter and interval between rock-bolts have been studied. It was found that: (1) Rock-bolt has a notable effect on the stability of the surrounding rocks. Increasing the diameter of rock-bolts contributes to significantly reduce the plastic area and the vertical displacement. With the increase of the quantity and the diameter of anchors, the degree of influence gradually weakens, showing the logarithmic relationship; (2) When the quantity and diameter of rock-bolts are constant, in the case in which the intervals between rock-bolts are various, the plastic area of surrounding rocks and the axial stress of rock-bolts are larger than that of the same distance, while the vertical displacement are uniform. (3) The excavation process has a great influence on the axial stress of rock-bolts and the redistribution of stresses in the surrounding rock.
Keyword: BIM, Anchoring system, Rock-bolts

1 Introduction

The equilibrium state of stress in the surrounding rock will be destroyed owing to excavation. For the surrounding rock of the affect region, a large amount of energy is going to be released by coordinate deformation as well as stress redistribution. If the redistribution stress exceeds the surrounding rock's strength, a lot of new cracks will occur

* Corresponding author : wzchen@whrsm.ac.cn.

and at the same time, the existed cracks will extend. Finally, all of them affect each other leading to instability and destruction.

Anchoring system can transfer and enhance the strength and self-stability of the rock body by being embedded into it. Scholars did a plenty of works on the anchoring mechanism and optimal design of the anchors and concluded the anchor mechanism of suspension theory, supporting theory, composite theory, enhancement theory and pin theory [1,2]. However, the rock bolting design is still empirical or semi-empirical in practical engineering, due to the lack of understanding of bolt reinforcement mechanism [3,4], and the effects and contributions of several parameters are not very clear[5]. Ideally, large-scale physical experiments should be carried out to study the effects and contributions of these parameters. Moreover, (1) it seems to be difficult to get a consistent result, owing to the limitation of the laboratory experimental conditions. For example, the size of the test body can't be cast large enough to satisfy the requirement of testing or to simulate the practical underground geological conditions[6,7]; (2) it is not very convenient for the test bodies to accommodate so many rock-bolts; (3) for larger diameter rock-bolt tests, there was always a lack of sufficient confinement to prevent the split of the test bodies under loading[8]. While, numerical simulations have the advantages of that (1) it is conceivable to satisfy the actual geological conditions as accurate as we need; (2) it can provide further insights and several significant improvements to these models including consistent testing environment which avoids issues of unbalanced forces and insufficient confinement; (3) it is reliable to examine every influence factors whilst obtaining reasonable results[8]; (4) it has the characteristics of repeatability, adjustability, convenience, accuracy, low cost and strong operability[9]. So numerical simulation is a good way to study the optimization of anchoring parameters. On the other hand, whether to generate numerical model accurately is an important factor affecting the accuracy of analysis results, and it seems to be difficult to simulate anchors of different design parameters quickly, accurately and parametrically. CATIA provides a way to build models by parameterizations automatically to reach the purpose of accurate and quick modelling.

In this paper, the simulation of excavation and rock-bolt reinforcement in a large scale of rock body has been performed by parameterized modeling in CATIA and analyzing in ABAQUS. It aims to further understanding of the influence of relevant parameters, the distribution of stress and stability of surrounding rocks for the optimization of anchors.

2 Model building and cases study

2.1 Model building

The parameters of the practical project including intervals and diameters of each anchor are easy to adjust for various research purpose in order to optimize the anchoring system by the means of parameterized design module of CATIA[10,11], so as to build the design model (Fig.1). The model is imported into ABAQUS for the CATIA-ABAQUS simulation model to analyze the influence of excavation on the stability and stress redistribution of surrounding rock, and investigate the factors and contributions of the influence factors on the stability of surrounding rock to realize the fusion of numerical simulation and BIM. The simulation process is shown in Fig.2. For the sake of simplification, the geological-terrain model was simplified as a cube in this research.

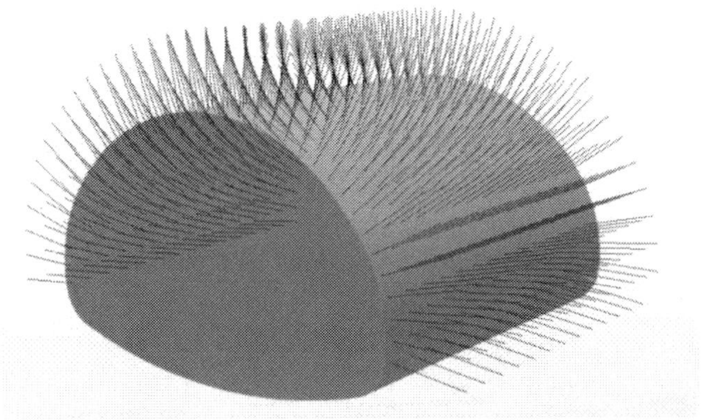

Fig. 1. The layout of the model

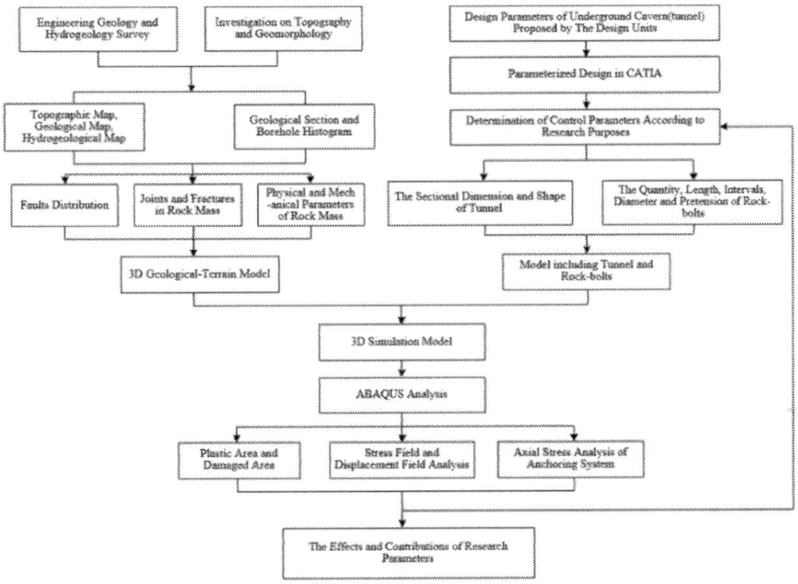

Fig. 2. Simulation process

The study model consisted of two types of materials: rock body and steel (rock-bolt). In the numerical simulation, the rock body was modelled as the Mohr-Coulomb criterion, elasto-plastically and the rock-bolt was modelled as truss of linear elastic model. The interface between rock-bolts and rock mass was simulated by embedded elements which means the nodes of rock-bolt elements are inserted in the rock mass element by interpolation method. Reasonable material property is crucial to obtain correct simulation results. Table 1 and 2 show the material properties that were used in the numerical simulation.

Table 1. Material properties of the rock body

Density/(kg · m^{-3})	**Elastic modulus/MPa**	**Poisson's ratio**	**Cohesion/MPa**	**Friction angle/(°)**
2500	1200	0.35	0.2	25

Table 2. Material properties of rock-bolt

Density /(kg ·	**Elastic modulus**	**Poisson's**	**Coefficient of expa**	**Pretension/M**

m^{-3})	/MPa	ratio	nsion	Pa
6000	200000	0.25	1.3×10^{-5}	260

2.2 Cases study

To avoid steel wasting and to shorten the period of construction, it's important to optimize the design of rock-bolts, including the length and the diameter of rock-bolts, rock-bolt inclination, interval between rock-bolts, and so on. So numerical simulation was carried out to investigate the influences and contributions of various parameters on the stability of rock body, stress distribution of rock body and rock-bolts, and the convergence of surrounding rock.

Rock-bolts with different setting formations and diameters, 18, 22, 25, 28, and 32mm, were embedded in rock body which was with the size of 260m×20m×154m. The length of rock-bolts was 4m, whole of which was embedded into the rock body. The boundary conditions applied in the model were that the bottom side of the model was displacement limited at 3 directions, and the side face was limited to normal displacement constraint. Every two layers of rock-bolts were set as rectangular (Fig.1). In this paper, we considered the length and the inclination of rock-bolt as invariants, and the variables investigated are listed in Table 3. Besides, this paper considered the effect of excavation process[12]by dividing the tunnel into three sections for full section excavation.

Table 3. The variables and invariants investigated in this paper

Variables		**Invariants**
Intervals between rock-bolts	1.6m×1.6m 1.2m×1.2m 0.8m×0.8m 0.6m×0.6m 0.4m×0.4m	90°, 4000 mm length, 260MPa pretension, rock body strength
Rock-bolt diameter	18mm, 22mm, 25mm, 28mm, 32mm	

3 Numerical analysis and results

As the quantity of output results were so large, only the main results were presented here. We considered the convergence to be one of the crucial criterion for the stability of the tunnel. So this paper discussed how the quantity (interval between rock-bolts longitudinal and tangent) and the diameter of rock-bolts affect the vertical displacement of the tunnel. Additionally, the characteristics of plastic zone, stress field and the axial stress of rock-bolts were studied.

3.1 Plastic zone analysis

Only part of typical characteristics of the plastic zones were discussed in which no rock-bolt existed or the distance between rock-bolts longitudinal and tangent are both 1.6m, 0.8m, 0.4m, respectively. Fig.3 shows that the plastic zone without rock-bolt is mainly concentrated on the arch hance and arch spandrel, like a couple of wing. While the features changed when rock-bolt existed and different characteristics could be seen as the distance between rock-bolts differs. The area of plastic zone reduces as the setting density becoming large and focuses on the bottom of the arch which means rock-bolt is useful for inhibiting the extension of stress and strengthening the surrounding rock to improve its own strength.

In fact, what rock-bolts do to improve the stability of the surrounding rock is called anchorage effect. Excavation of tunnel leading to unload of rock and stress redistribution is the key factor to rock deformation. To avoid exceeding displacement, anchoring system is needed. As long as the surrounding rock deforms, the rock-bolts embedded in the rock body will lead to additional radial force and the stress distribution of the rock translates from two-directional stress state to three-directional stress state[13]. The tunnel whose section is circle forms a supporting ring and the tunnel with arch section forms a supporting arch, both of which avoid the deterioration of the property of rock body. Additionally, the physical and mechanical properties of the medium composed of steel and rock are ideally much higher than rock. In this paper, we improved the properties Elastic module, Poisson's ratio, Cohesion by 10%[14].

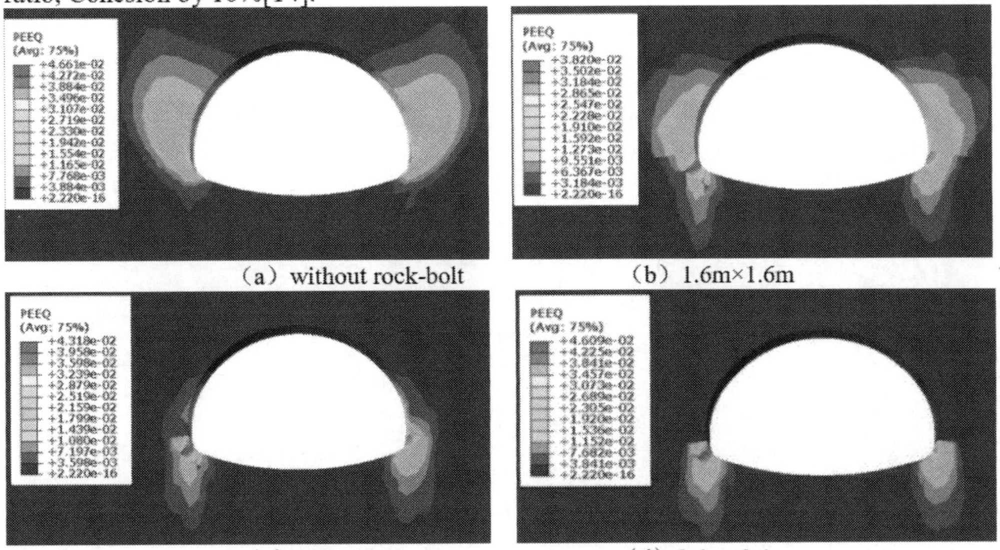

(a) without rock-bolt (b) 1.6m×1.6m
(c) 0.8m×0.8m (d) 0.4m×0.4m

Fig. 3. The plastic zone distribution with setting density increasing

3.2 Stress field analysis

Take the stress distribution of the tunnel vault (called "line" below) with various rock-bolt's setting formations for example.

Fig.4(a) shows that the regular of the radial stress distribution without rock-bolt is different from that with rock-bolt. If no rock-bolt is embedded, with an increase of the distance from the tunnel wall, the radial stress of "line" increases gradually, then it's going to be stable. When rock-bolts exist, the radial stress of "line" increases to a maximum and then decreases gradually in the anchoring area. Beyond the anchoring area, the radial stress increases to a stable value. Take the case in which the distance between longitudinal and tangent rock-bolt are both 0.6m for example, the radial stress of the first observation point on the tunnel wall is 0.6MPa, while it increases with the distance from the tunnel wall, and reaches the maximum at the fourth observation point. Then, it decreases to 0.28MPa at the fifth observation point. At last, out of the anchoring area, the radial stress raises as the distance from the tunnel wall. What's more, the more the rock-bolts are, the larger the radial stress on the tunnel wall is, and the maximum of the radial stress increases as well. When the distance of longitudinal and tangent rock-bolts are both 1.6m, 1.2m, 0.8m, 0.6m, and 0.4m, the radial stress on the tunnel wall is 0.167MPa, 0.207MPa, 0.398MPa, 0.604MPa, 1.006MPa, respectively; the maximum stress of this line is 0.247MPa, 0.314MPa, 0.483MPa, 0.659MPa, 1.062MPa, respectively. The mechanism of this phenomenon is that excavation led to rock low stability and stress redistribution. If the anchoring system is not applied, the strength of the rock decreases to the residual stress, instead, the stress of the surrounding rock redistributes again because of the anchoring force which constitutes of the stress of the surrounding rock with initial stress. As a result, the maximum stress and residual stress increase as the strength of the anchoring system improves. On the other hand, due to the enhancement theory of rock-bolt, the physical and mechanical properties of the rock-bolt are improved and they improve a lot as the setting density is enhanced which also leads to the improving of the anchoring effect, suggesting that the radial stress improves with the setting density.

As shown in Fig.4(b), no matter the rock-bolt exists or not, the regular of the tangent stress on the vault of the tunnel exists a concentration value in the anchoring area, and then decreases to a stable value. But the difference is that the maximum of the tangent stress anchored is much bigger than that without rock-bolt. The tangent stress and the maximum tangent value are both increase with the setting density increasing, and they tend to be the same value at last. For example, when the longitudinal and tangent interval are both 1.6m, 1.2m, 0.8m, 0.6m, and 0.4m, the tangent stress of the tunnel wall is 0.698MPa, 0.729MPa, 0.815MPa, 0.871MPa, 0.952MPa, respectively; the maximum value of the tunnel vault is 0.792MPa, 0.853MPa, 0.972MPa, 1.061MPa, 1.212MPa， which is much bigger than that without rock-bolt(0.60MPa, 0.61MPa).

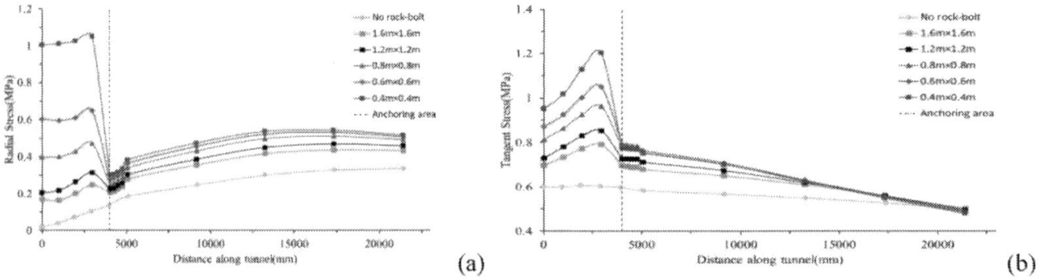

Fig. 4. The distribution of radial and tangent stress at various cases. (a) Radial stress, (b) Tangent stress.

3.3 Displacement field analysis

3.3.1 Influence of quantity

The effect of different setting density of rock-bolts on the radial convergence of the tunnel vault is shown in Fig.5. It can be seen that the stability of the surrounding rock has been greatly improved. Take the rock-bolts set as 1.6m × 1.6m for example, the vertical displacement of the vault is about -26mm, which reduces by 23.5% compared with -34mm without rock-bolt, and reduces by 27.6%, 35.3%, 39.7%, 44.1%, respectively, for rock-bolts set as 1.2m × 1.2m, 0.8m × 0.8m, 0.6m × 0.6m, 0.4m × 0.4m. With the longitudinal and tangent distance becoming small, the vertical displacement of the tunnel vault reduces significantly, the degree of which is in terms of the quantity of rock-bolt (Table 4). As shown in Fig.6, the deformation decreases as the quantity of rock-bolts increases, suggesting that they have obviously positive correlation with each other. When the quantity of rock-bolt is not very large, the deformation increases significantly with it and it does not increases obviously as the quantity increases, suggesting the property of convex function [9]. The characteristic is consistent with the results obtained by DEM[15]. It indexes that the regular of "the more, the better" is not reliable, we'd better choose the proper setting formation according to the vertical displacement.

Table 4. The quantity of anchors of different setting formation

Formation	1.6m×1.6m	1.2m×1.2m	0.8m×0.8m	0.6m×0.6m	0.4m×0.4m	1.0m×1.4m	0.6m×1.1m

Quantity	286	522	1276	1972	4472	525	522

The effect of the quantity of rock-bolts on the vertical displacement has been studied by the means of statistical method. The expression about the quantity of rock-bolts and the vertical displacement for tunnel vault and spandrel are listed in Eq.(1) and Eq.(2), respectively, and these equations can be translated into ordinary formation (Eq.(3)).

$$y_N=2.897\times ln(x_N)-42.7 \tag{1}$$

$$y_N=3.307\times ln(x_N)-43.403 \tag{2}$$

$$y_N=a_N\times ln(x_N)+b_N \tag{3}$$

where y_N is the vertical displacement, x_N is the quantity of rock-bolts. In these equations the R^2 is 0.9882 and 0.9827, respectively, which means that the statistics fit well.

Obviously, there is logarithmic relationship between the quantity of rock-bolts and the vertical displacement. In Eq.(3) a_N and b_N are parameters needed to be confirmed by the means of experiment or numerical simulation which are in terms of the lithology. As long as the expression is quantified, it is possible to get the quantity of rock-bolts according to the conditional parameters -- vertical displacement, then to confirm the setting formation.

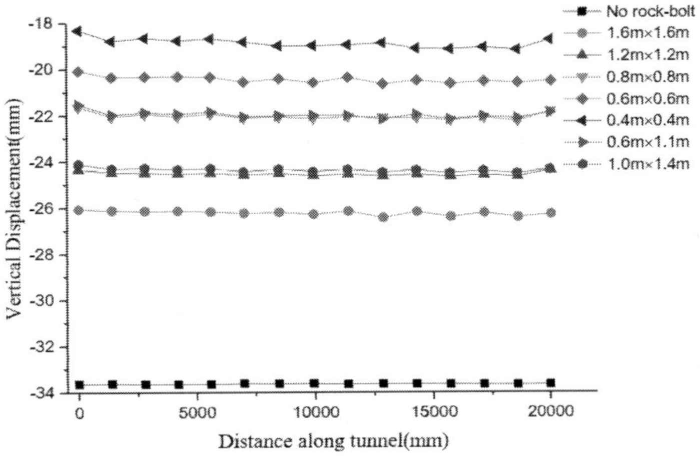

Fig. 5. The vertical displacement of the tunnel vault at different cases

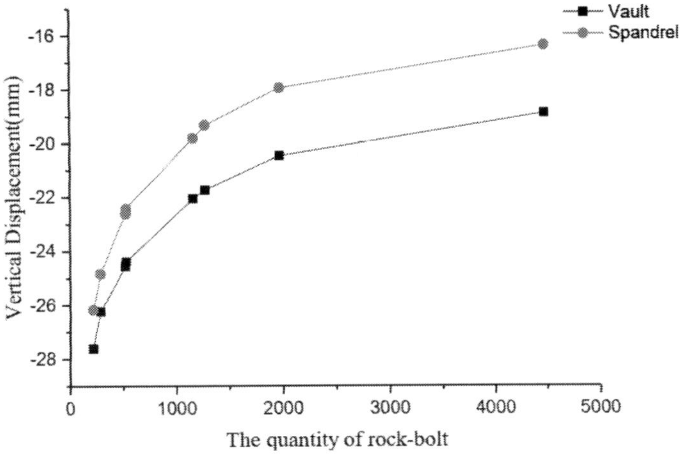

Fig. 6. Influence of the quantity of rock-bolts on the vertical displacement

3.2.2 Influence of diameter

In this numerical model, the longitudinal and tangent distance of rock-bolts are both 0.6m, and the diameter is set to 18mm, 22mm, 25mm, 28mm, and 32mm, while other parameters are kept consistent with those in Table 1, 2. The results of the vertical displacement are shown in Table 5, and the features of the vertical displacement versus the diameter of rock-bolts are shown in Fig.7, which is similar with the effect caused by the quantity of rock-bolts. With the diameter of rock-bolt increasing, the vertical displacement does not reduce significantly. Similarly, we studied it with the method of statistical method, and the result is concluded with Eq.(4) and Eq.(5) for tunnel vault and spandrel, respectively. Like the phenomenon discussed above, the parameters called a_D and b_D are parameters which should be defined by laboratory test, field test or numerical simulation. So for the specifical project, the diameter of the rock-bolts can been chosen using Eq.(6).

$$y_D=4.8125\times ln(x_D)-36.011 \tag{4}$$

$$y_D=4.77\times ln(x_D)-35.609 \tag{5}$$

$$y_D=a_D\times ln(x_D)+b_D \tag{6}$$

where y_D is the vertical displacement, x_D is the diameter of rock-bolts. In these equations the R^2 is 0.9953 and 0.9945, which means that the statistics fit well.

When the rock is unloaded, the surrounding rock deforms into blocks and the radial anchoring force is imposed on the rock body leading to rock compression. The cracks are closed, then the cohesion and frictional angle increase, the strength of the surrounding rock is enhanced[16]. If the quantity of the rock-bolts reaches to a critical value, the cracks are all closed and the rock body reaches to a limitation of the extrusion deformation. Beyond the critical, new cracks will appear which weaken the integrity and the stability of the rock.

Table 5. The vertical displacement under various diameters of rock-bolts

Diameter（mm）	15	20	25	30	35
Vertical displacement（mm）	-23.325	-21.573	-20.455	-19.665	-19.069

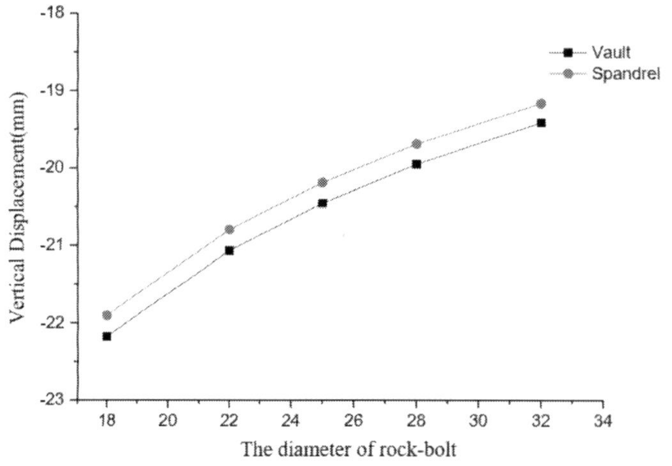

Fig.7. The vertical displacement with different diameters

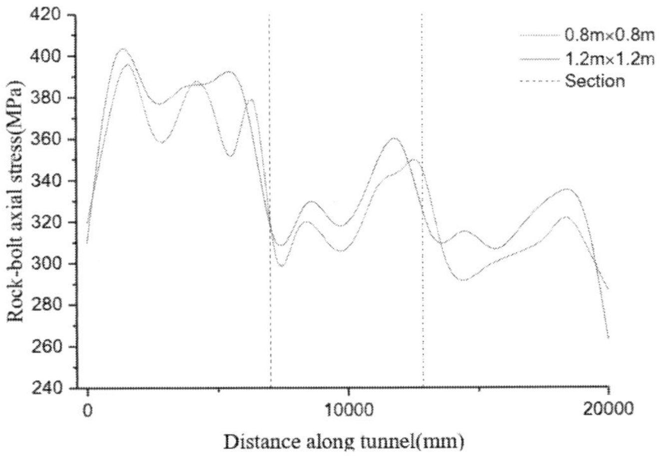

Fig. 8. Influence of excavation process on axial stress

3.4 Axial stress analysis

From the results discussed above, the effect on the stability of the surrounding rock of the longitudinal and tangent interval is due to the quantity of rock-bolts, suggesting that it is none of the interval's business, but it will differ a lot for the axial stress of rock-bolt.

When the quantity of rock-bolt is closed to each other, we need to study the axial stress distribution of rock-bolt. Take the cases in which rock-bolts are set as 1.0m×1.4m, 1.2m×1.2m for example, there are 525 rock-bolts in the former case, and 522 in the latter one.

The distribution of rock-bolt axial stress versus each excavation section is shown in Fig.8, and it can be seen that the axial force of the rock-bolt is a fluctuating distribution in each excavation section, with multiple extremums and one maximum. In order to do some research on the distribution of rock-bolt axial stress along the length, we studied the average axial stress of the tunnel vault and the hance in two cases. Fig.9(a) shows the characteristics of the axial stress versus the length of rock-bolt in the hance of the tunnel, showing a maximum at 601.92mm with value of 830MPa for the case of 1.0m×1.4m, and a maximum at 401.25mm with value of 828.6MPa for the case of 1.2m×1.2m. It can be seen that the maximums are similar in the two cases, but the maximum stress of the latter is closer to the tunnel wall than the former, and the maximum stress of the rock-bolt in both formations are offset the middle position and close to the inside part of the tunnel.

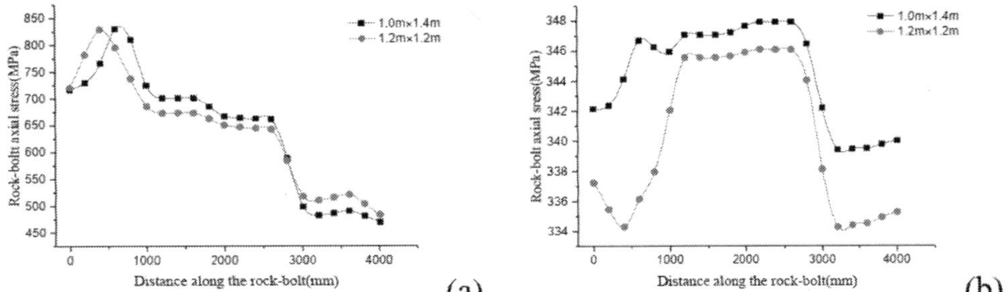

Fig. 9. The axial stress distribution along the length of rock-bolt under different setting arrangements. (a) The hance of tunnel, (b) The tunnel vault.

Fig.9(b) shows the distribution of axial stress at the tunnel vault in the two cases, presenting the similar features, but the axial stress of the former is slightly higher than that of the latter.

By analyzing the distribution characteristics of the axial stress of the two cases, it can be seen that when the quantity of rock-bolts is similar, the effect they do on the stability of the surrounding rock is also the same and the vertical displacement can be calculated by Eq.(3) or Eq.(6). While, the plastic area of the former case is much larger than that of the latter and the axial stress of rock-bolts is also higher which means that the risk of instability increases. Therefore, we'd better choose the equivalent longitudinal and tangent intervals between rock-bolts when the quantity of rock-bolts is constant.

4 Conclusions

In this study, an accurate model was built by CATIA, and was employed in the mechanical analysis of the bolt-reinforced tunnel which was excavated by full excavation for three sections. Besides, the factors and contributions of the parameters which have influence on the rock body have been studied by the means of numerical simulation. Primary conclusions are summarized as follows.

(1) The increasing quantity of rock-bolts contributed to significantly reduce the plastic zone and the vertical displacement. While, it does not increase obviously any more with the number of rock-bolts increasing, which is the law of logarithm function, so as to the effect of the diameter of rock-bolt.

(2) The process of excavation matters a lot to the rock-bolts axial force, and several extremum values and one maximum exist on the rock-bolt axial force along the length of tunnel.

(3) For the rectangular arrangement formation, the influence of the intervals between rock-bolts on the stability of the surrounding rock is due to the impact of the quantity of anchors, but the longitudinal and tangent rock-bolts with same intervals is stable than that of different one.

References

1. S.R. Zhang, Y. Gu, Z.L. Zhang. The optimized design of rock-bolts supporting the large-scale underground cavities[J]. Journal of Hydroelectric Engineering, **26(5)**:47-52(2007)
2. L.W Zhang, R. Wang. Research on status quo of anchorage theory of rock and soil[J]. Rock and Soil Mechanics, **23(5)**:627-631(2002)
3. Y. Cai, T. Esaki, Y.J. Jiang. An analytical model to predict axial load in grouted rock bolt for soft rock tunnelling[J]. Tunnelling and Underground Space Technology, **19(6)** : 607-618(2004a)
4. Y. Cai, T. Esaki, Y.J. Jiang. A rock bolt and rock mass interaction model[J]. International Journal of Rock Mechanics & Mining Sciences, **41(7)**:1055-1067(2004b)
5. J.C. Gu, J. Shen, A.M. Chen, et al. Model testing study of strain distribution regularity in rock mass caused by prestressed anchorage cable[J]. Rock Mechanics and Rock Engineering, **19(Z1)**:917-921(2000)
6. A.T. Haile. A mechanistic evaluation and design of tunnel support systems for deep level South African mines. PhD thesis (unpublished), University of Natal, Durban, South Africa(1999a)
7. A.T. Haile. Observation of the dynamic support performance of South African tunnel support systems. In: Proceedings rock support and reinforcement practice in mining,

Kalgoorlie, Australia, 335–341(1999b)
8. L. Li. Parametric Study of Rockbolt Shear Behaviour by Double Shear Test[J]. Rock Mechanics and Rock Engineering, **49(12)**:4787-4797(2016)
9. Q.L. Dong, Research on deformation law of weak broken roadway and optimization of bolt support parameters under mining influence[D]. GuiZhou University(2018)
10. X.C. Han. Research on Application of BIM Concept Tunnel Portal Design Program[D]. Chongqing Jiaotong University(2018)
11. X. Qin. Research on Railway Geological Route Selection Method Based on BIM[D]. Shijiazhuang Tiedao University(2018)
12. G.L. Han. Analysis of Mechanism of Bolt-Anchor Cable Combined Support for Weak Rock Cavern by ABAQUS[J]. Metal Mine, 75-79(2008)
13. P.P. Chen, Q.W. Xu, H.H. Zhu, et al. Model experiment study on the failure characteristics of weak surrounding rock mass above tunnel arch and the anchoring effect of bolts[J]. Chinese Journal of Rock Mechanics & Engineering, **35(S2)**:3561-3569(2016)
14. H. Chen, W.Z. Ren, D. Li, et al. Numerical simulation and model test study of mechanism of bolt in deep tunnel[J]. Rock and Soil Mechanics, **32(S1)**:719-724(2011)
15. H.N. Wang, X. Guo, M.J. Jiang, et al. Investigation of rock bolting for deeply buried tunnels via a new efficient hybrid DEM-Analytical model[J]. Tunnelling and Underground Space Technology, **82**:366-379(2018)
16. H. Chen, W.Z. Ren, Z.G. Shu, et al. Model test study and numerical analysis of mechanism of anchor bolt under different supporting conditions[J]. Rock & Soil Mechanics, **33(S1)**:277-282(2012)

An integrated approach for jointed rock slope stability analysis using stochastic discrete fractures: A case study in mountains area

Weizhong Chen[1,3,*], *Luyu* Wang[1,2], *Xuyan* Tan[1,2], and *Jiang* Lei[1,2]

[1] State Key Laboratory of Geomechanics and Geotechnical Engineering, Institute of Rock and Soil Mechanics, Chinese Academy of Sciences, Wuhan 430071, China

[2] University of Chinese Academy of Sciences, Beijing 100049, China

[3] Research Center of Geotechnical and Structural Engineering, Shandong University, Jinan 250100, China

Abstract. Stochastic discrete fractures are ubiquitous in practicing engineering and contribute much to the stability for geological mediums. In contrast to the traditional treatments based on continuum mechanics, we propose an integrated approach using discrete fractures based on an improved hybrid finite element method (FEM). A typical jointed rock slop is selected as a case study and then was simulated in the light of geological conditions for the study site. The basic equations of fracture domains are derived to generate stochastic discrete fractures which render a fundamental to the model. To implement this procedure through computer program, discrete fractures generating algorithm was developed. In the aspect of discretization, we presented a meshing scheme and proposed an algorithm to create fracture elements. To evaluate rock slope stability in a quantitative way, safety factor is calculated by strength reduction method. Numerical simulations were carried out showing the mechanical behaviors of jointed rock slope under different boundary conditions.

1. Assement of the stability for a jointed rock slope

1.1 Role of the slope stablilty analysis in mountain area

The mechanical behaviors of fractured mediums, for instance, jointed rock slope, weathered geological body and fractured reservoirs, are mainly controlled by mechanical and hydraulic properties of discrete fractures, which is one of the most typical types of discontinuous surfaces. Discontinuous in a natural geological structure include fractures, faults, cleavages and weak bedding planes, which lead to the anisotropic and homogeneous for geological mediums [1–3]. In especial, for a jointed rock slope, the influences of fractures on slope stability and mechanical response is vital important to engineering design. In the mountain area, there are a lot of infrastructure projects built on fractured geological mediums, such as the highway roadbed, bridge foundation and underground excavation. Stochastic discrete fractures are ubiquitous in this type of region and contribute much to the stability for a geological medium [4,5]. Apart from field investigation and experiments, it is essential developing an effective numerical approach to study the stability of fractured rocks. However, numerical simulation for a fractured medium is still a challenge topic in the field of computational mechanics [1,6,7,8]. In this paper, we mainly focus on the simulation approach about jointed rock slope stability.

1.2 Traditional approaches in slope stability analysis

In the aspect of slope stability analysis, a lot of literatures have been reported different approaches to study it [9–15]. However, it is essentially different that mechanical properties and geological structure patterns under the situations of soil and rock slope. The traditional treatment for soil slope is mainly on continuum mechanics combining with limit equilibrium method (LEM) [16,17]. In order to evaluate slope stability in a mathematical principle, Zienkiewicz introduced an index in his famous article [18], the well-known strength reduction factor (SRF), to show the stability ability for a soil slope. In the recent decades, the strength reduction method (SRM) has been successful used in practice and engineering design [19,20]. Nevertheless, this method is not very effective for a jointed rock slope. The main factor leading to this problem is that the effects of fractures on slope stability were neglected. To overcome the shortcoming of SRM, Zienkiewicz proposed an equivalent continuous medium model [21] to deal with it, in which he presented a novel

*Corresponding author: wzchen@whrsm.ac.cn (Weizhong Chen)

luyu.wang@hotmail.com (Luyu Wang) tanxuyan16@mails.ucas.ac.cn (Xuyan Tan) leijiang14@mails.ucas.ac.cn (Jiang Lei) finite element material, called jointed material, to reflect the impacts of fractures. But for a sparse fracture medium, there are still many unsolved problems for the method. On the basis of his theory, a lot of literatures have been published improved approach to simulate fractured rock slope stability [22–25], but they are all based on continuum mechanics. To simulate the influences of fracture on rock slope stability, many novel approaches were proposed, for instance, distinct element method [26,27], numerical manifold method [28], discontinuous deformation analysis [29], hybrid finite discrete method [30], mesh free analysis [31,32]. To integrate fractures into numerical model, these above methods all mainly based on an important conceptual model, discrete fracture network (DFN), which is one of the most popular treatments to deal with stochastic fractures [33–37].

1.3 Jointed rock slope stability analysis considering the random fractures

As mentioned above, in general, researchers studied fractured rocks through DFN. Mechanical and hydraulic characteristics of DFN are important to problems of petroleum development, fractured reservoirs, wastes disposal, hydrogeology, geothermal development [1,6]. Field investigation results demonstrates the distribution of fractures in a network is complex and random, and is treated usually by DFN. The concept of DFN was first introduced by Long et al. [38]. In the aspect of applications of DFN, Lei et al. [6] published a review article to state the use of DFN for geomechanical-hydrological process in fractured mediums. They emphasized that the coupled problems of DFN model is vital important to the fundamental of rock mechanics. To implement the DFN by computer programs, the corresponding computational algorithms should be developed. But there are still unsolved issues related to this topic, such as mesh generation problem, three-dimensional modeling and visualization, simulation techniques of flow and coupled process.

Following the previous works by our group [40], the present study performed a numerical investigation for a jointed rock slope in the light of the geological conditions of a case study. An integrated approach using discrete fracture network and based on the improved hybrid FEM was proposed. In section 2, computer algorithms were proposed to generate a DFN and to deal with the grid problem for the study site. Combining with SRM, the improved hybrid FEM was presented in section 3. Then, numerical simulation was carried out to study slope stability, and different methods were employed to show the differences between intact and fractured rock slope.

2. Treatment of the stochastic discrete fractures network

To simulate the mechanical behaviors of discrete fractures networks (DFNs) including bedding planes and random fractures which exist in jointed rock slope, an improved hybrid finite element method was employed in this study. In such approach, the effects of DFNs on fractured medium stability were considered through integrated into finite element method code. It is notable that this type of method was developed by our team [40] and also has been employed in the stability analysis of underground excavation.

2.1 Generation of the stochastic discrete fractures

The generation of DFNs is vital important to simulation of fractured mediums. In particular, the characteristics of a DFN need to be described a prior so that it can be generated. In practice, discrete fracture distribution often treated by following a certain statistical law. In this means, there are a lot of research works on fracture networks distributions and have been widely reported in literatures [2,4–6].

Based on the above disscussiones, a stochastics discrete fracture generation algorithm was developed. According to this algorithm, we implemented it by MATLAB and obtained the model. The configuration of the jointed rock slope in our study site, Nayong bank side, is illustrated in Fig. 1, which considered the effects both of certain bedding planes and random fractures on slope stability. The model range and fracture patterns were created in the light of field investigation. To avoid the boundary effects, slope model boundaries are used with an intact rock buffer zone between the fracture network and model boundary. This treatment of boundary effects also has been employed by Bisdom et al. [39].

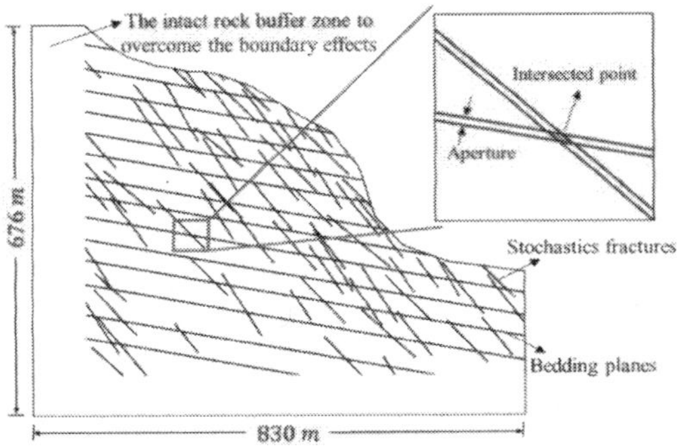

Fig. 1. Jointed rock slope numerical model considering bedding planes and stochastic fractures

2.2 Discretization for the jointed rock slope model

The DFN model is vital important to the generation of finite element grids. In the situation of a complicated fracture pattern, the generation of a high-quality mesh is difficult due to the random partition configuration of the rock matrix. In addition, the discrete fractures are ubiquitous and possess a thin aperture resulting in a problem in the mesh partition process. To deal with it, a lot of literatures reported different method to carry out it. In this section, we improved a novel strategy to discrete the DFN model.

For the current model, jointed rock slope, to simulate the mechanical behavior and the aperture variation of fractures, the discrete fractures and matrix were discretized by quadrilateral elements and triangular elements, respectively. An algorithm has been developed to generate the fracture elements and the intersected elements for the discrete fractured slope model. The completed grid using fracture element generation algorithm is displayed in Fig. 2, in which the red lines indicate fracture elements.

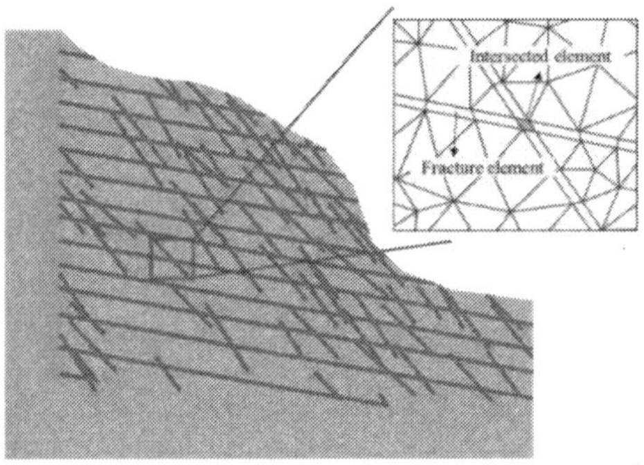

Fig. 2. Completed grid using fracture element generation algorithm and the intersected element between two fractures

2.3 Geomechanical model for discrete fractures and rock matrix

To describe the mechanical behaviors of stochastics fractures and bedding planes to exhibit its weakening and discontinuous effects, fracture elements were employed to deal with it. In another aspect, the rock matrix would show different response with that of fractures, thus the constitutive relations based on traditional continuum mechanics were introduced.

The fractures were modeled by the fracture elements. According to that, we have been defined the different stages for the failure conditions of fracture elements. To explain it, two key points to implement the simulation are shown as follows.

1) Elastic stage. When the normal and shear displacement, denoted by δ_s, δ_n, of fracture element are less than the critical displacement in these two directions, denoted by δ_s^0, δ_n^0, the mechanical responses of fracture element show elastic property, which are given by

$$\begin{pmatrix} \sigma_n \\ \sigma_s \end{pmatrix} = \begin{pmatrix} K_{nn} & K_{ns} \\ K_{sn} & K_{ss} \end{pmatrix} \begin{pmatrix} \delta_n \\ \delta_s \end{pmatrix} \tag{1}$$

Where σ_n and σ_s are normal and shear stress corresponding to δ_n and δ_s, respectively. K_{nn} and K_{ss} are pure normal and shear stiffness, respectively. K_{ns} and K_{sn} are the coupling stiffnesses to describe the interaction between normal and shear loads, which can be neglected based on the results by [40].

2) Damage stage. In the light of above discussions, when δ_n and δ_s are greater than δ_s^0, δ_n^0, the mechanical parameters of fracture element would reduce and the effects of damage should be considered into constitutive relation.

$$\begin{pmatrix} \sigma_n \\ \sigma_s \end{pmatrix} = (1-D) \begin{pmatrix} K_{nn} & K_{ns} \\ K_{sn} & K_{ss} \end{pmatrix} \begin{pmatrix} \delta_n \\ \delta_s \end{pmatrix} \tag{2}$$

Where D is the damage factor introduced by [54]. Based on the solving algorithm in the present work, the formula of D is defined by

$$D = \frac{\delta_m^f (\delta_m^{\max} - \delta_m^0)}{\delta_m^{\max} (\delta_m^f - \delta_m^0)} \tag{3}$$

Where δ_m is effective displacement to describe the effects of combination of δ_n and δ_s. $\delta_m^{\max}$ is the effective displacement threshold when the damage start. δ_m^0 and δ_m^f are the effective displacement at peak and final failure of fracture element, respectively.

To decide the condition that when does the damage process occur, a criterion was used in our model, which is proposed by [41], and the peak values of stress are denoted as σ_n^0 and σ_s^0

$$\left(\frac{\langle\sigma_n\rangle}{\sigma_n^0}\right)^2+\left(\frac{\sigma_s}{\sigma_s^0}\right)^2=1 \tag{4}$$

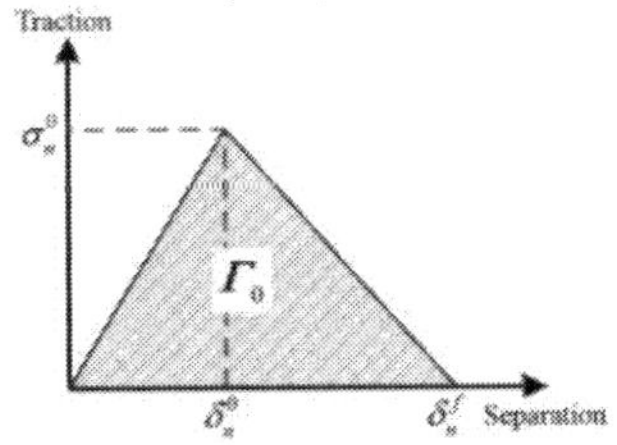

Fig. 3. Traction-separation law of fracture element during deformation

As shown in Fig. 3, the curve of fractue element during deformation is given [40], where the symbol Γ_0 denotes dissipated energy. The damage factor is motivated while the stress condition satisfying the above equation. Where the symbol $\langle\sigma_n\rangle$ indicates the $\sigma_n=0$ under compression condition.

3. An improved hybrid FEM approach combining with SRM

The above discussions render a fundamental to the hybrid finite element method improved by us. To simulate the fractured medium and to include the effects of discrete fractures on it, there are many works on this field [26–32], in which the hybrid FEM is one of the most widely used approach for the problem. Several types of the method have been proposed in the light of specific problems [30]. The general treatment of the improved method tackles the discontinuous surfaces and continuous matrix in a unified model. In our simulation, an important task is to calculate the safety factor (SF) of jointed rock slope. Based on the strength reduction method (SRM) proposed by Zienkiewicz [18], we integrated SRM into the hybrid FEM considering the impact of discrete fractures to obtain SF. The flowchart of the numerical approach for jointed rock slope is displayed in Fig. 4.

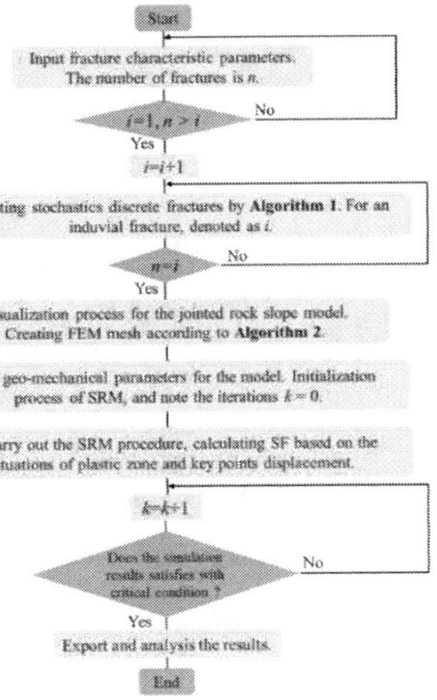

Fig. 4. Flowchart of the numerical approach for jointed rock slope

4. Numerical results and discussions

As mentioned above, a case study for a jointed rock slope was carried out using our improved hybrid FEM format based on the field investigation results for Nayong bank slope in Liupanshui city, Guizhou, China. The stochastic discrete fractures and bedding planes were considered by integrated the DFNs into the model. According to geological prospecting data, the overall stability condition is mainly affected by the complex fracture networks and the natural strati-graphic interfaces. In the following analysis, we focus on the mechanical behaviors of jointed rock slope under different boundary conditions, for instance, natural state and loading state. Another important intend for the simulation is to compare the differences between intact and fractured slopes to investigate the effects of discrete fractures on landslide and failure modes.

4.1 Geomechanical model for discrete fractures and rock matrix

To analyze the stability condition of the jointed rock slope under a natural wording situation, we setup a boundary condition, where the model only subjected to the gravity force (g = 9.8 m/s^2). It is essential to carry out the stress initialization step, during which movement of fracture surfaces and model boundaries were constrained. After finished stress initialization and displacement balance procedures, the displacement state was released to let the whole model equilibrium. The mechanical parameters used for simulation is listed in Table 1.

Table 1 Mechanical properties of rock for the study site

Elasticity modulus	Poisson's ratio	Frictional angle	Cohesion	Density
10.5 GPa	0.25	42°	1.8 MPa	2600 kg/m3

To obtain the safety factor (SF) using the improved hybrid FEM and strength reduction method (SRM), according to the flowchart in Fig. 4, we introduced the reduce safety factor (SRF) integrated into our numerical analysis. Until emerging an obvious plastic zone, so that a slip surface and potential landslide would be induced and resulting from mechanical parameter reduction (Fig. 5b). Focusing on the slip region, the distribution of displacement vectors corresponding to the region is illustrated in Fig. 5c. To analyze the displacement evolution for the key points during slip process, several monitoring sites in Fig. 5a were selected to illustrate it. By the way of SRM procedure, mechanical parameters such as strength force and friction angle, were reduced until forming an obvious slip surface, leading to the displacement variation of key points located in slip surface exhibiting a sudden change, Fig. 6.

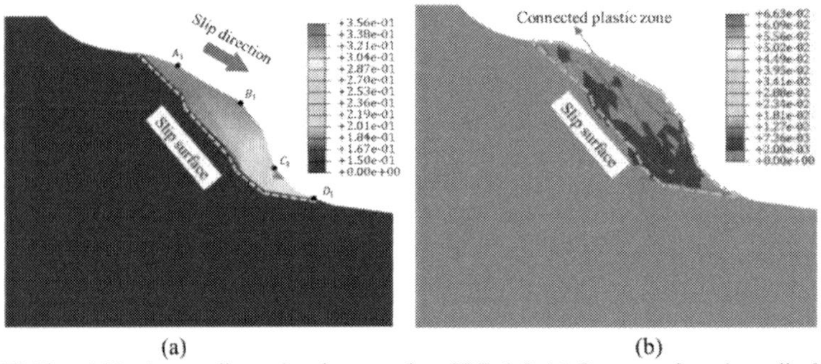

Fig. 5. Distributions of field variables in two-dimensional space when SRF=2.5. (a) Contour of resultant displacement, where the slip surface outlines the profile according to fracture pattern (unit: m). (b) Contour of equivalent plastic strain emerging a connected plastic zone caused by discrete fractures

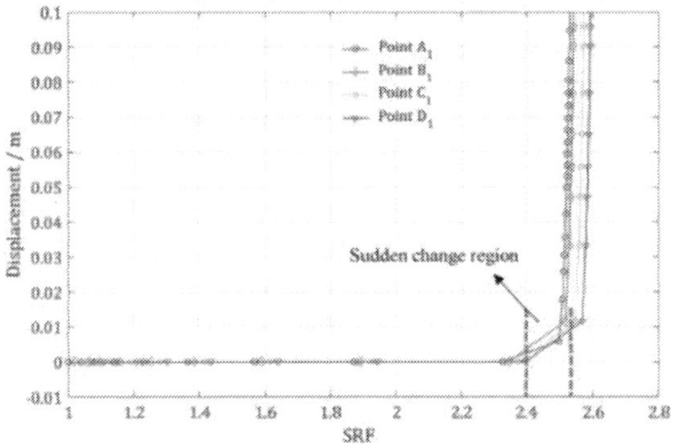

Fig. 6. Displacement evolution for key monitoring points during reduction process

4.2 The effects of bridge foundation loading on slope stability

It is essential to study the mechanical behaviors of the jointed rock slope during construction period, for which a large load would be applied to the slope surface induced by the bridge foundation. In the light to the bridge design plans, the foundation loads including both axial force p_f= 830578 kN and bending moment M_f= 1250283 kN·m, as shown in Fig. 7. From this figure, it is obvious that bridge foundation loads lead an influence region located near the position of foundation. On the other hand, comparing with the result of displacement contour displayed in Fig. 5a, the slip surface mode in the case of loading condition is significantly different from that of natural condition. The main factor which induced this difference is loading effects. It is reasonable to imply that the bridge foundation load would lead to the reduction of SF.

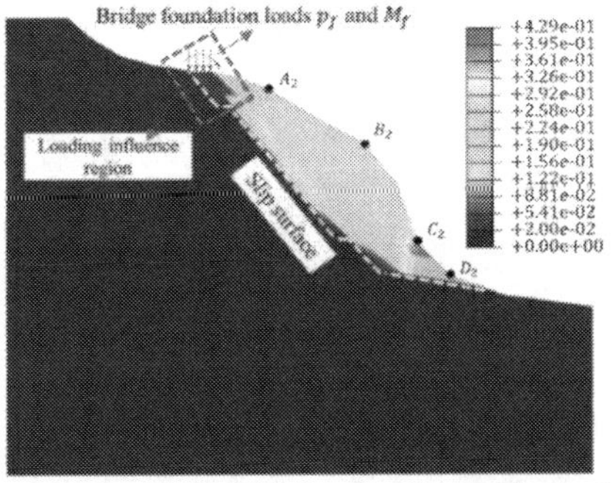

Fig. 7. Distributions of resultant displacement when SRF=2.42. Compared with Fig. 4a, the loading influence region indicates that the slip surface outlining effects of fracture pattern and bridge foundation loading (unit: m)

4.3 Sensitivity analysis for model parameters

To study the effects of different model parameters on the stability situations for the jointed rock slope, including fracture and rock matrix parameters, a sensitivity analysis was carried out to calculate the variation of safety factor (SF) under different mechanical parameters. Fig. 8 displays two sets of evolution curve between SF and material parameters under different conditions, which implies that both the parameters of discrete fractures and rock matrix have influence on slope stability. The parameters of friction angle and K_{in} reflect the mechanical properties of matrix and fractures, respectively. An increased magnitude of fracture parameter would induce a larger value of SF. It is obvious that the gradient variation of curves shown in Fig. 8a is smaller than that of in Fig. 8b, which also indicates the effects of fracture properties on slope stability are stronger than that of matrix. This conclusion shows the stability of a jointed rock slope is mainly controlled by discrete fractures and the attitude of stratum, where the effects of matrix property contribute a little proportion for that. [42] also has been studied the sensitivity analysis for model parameter of slope stability.

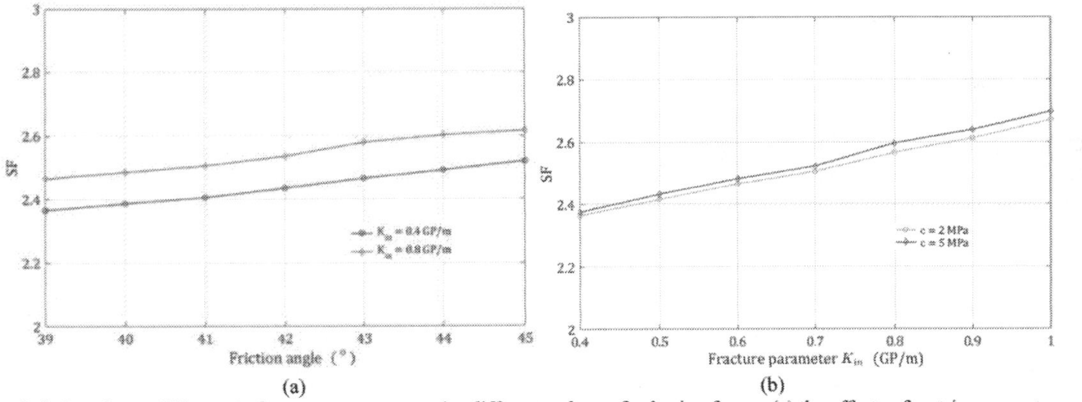

(a) (b)

Fig. 8. Safety factor (SF) versus fracture parameter under different values of cohesion forces. (a) the effects of matrix parameter on SF under different fracture parameter. (b) the effects of fracture parameter on SF under different cohesion force

4.4 Comparison analysis by different approaches: intact versus fractured rock slopes

To analyze the influences of stochastic discrete fractures on slope stability, we carried out a comparison analysis between these two different types of slope. Corresponding to the simulation results of loading condition shown in Fig. 5, the distributions of field variables (e.g. resultant displacement and equivalent plastic strain) are displayed in Fig. 9. There is another popular method to analyze the stability of slope, limit equilibrium method (LEM) [16]. The basic hypothesis of this method is to consider slope stability as an equilibrium problem. We employed a commercial code SLIDE to implement a LEM analysis in the light of conditions for the case study, as shown in Fig. 10. By the means of sliding surface search algorithm, there are many slip surfaces and all of them are circular, in which the minimum value of SF is 2.86. In general, the value of SF calculated by LEM is often great than that of obtained by our proposed method. This difference is mainly caused by not including the contributions of fractures. Because of the approach is based on the limit equilibrium theory, as a result, it cannot integrate a discrete fracture network into analysis procedure. A lot of improved suggests have been proposed to revise this problem, however, it still in the framework of equivalent approach, which is oversimplify for a natural fractured rock slope.

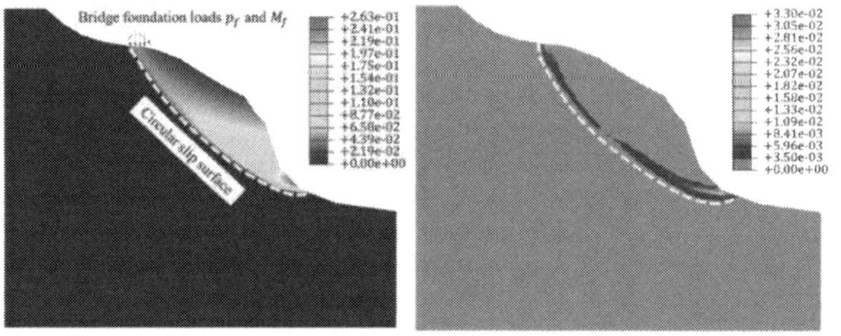

(a) (b)

Fig. 9. Distributions of field variables under loading condition when SRF=2.96. (a) Contour of resultant displacement, where the slip surface is circular (unit: m). (b) Contour of equivalent plastic strain

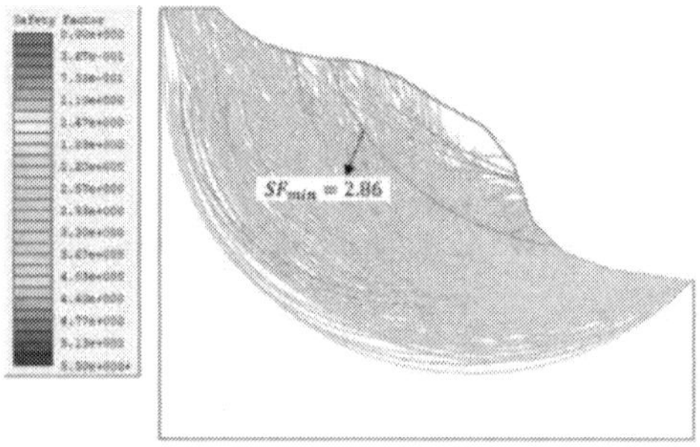

Fig. 10. Result of rock slope stability analysis by LEM analysis.fects of fracture pattern and bridge foundation loading (unit: m)

5. Conclusions

To investigate jointed rock slope stability, an integrated approach using discrete fracture network based on an improved hybrid FEM was proposed to study it. A typical jointed rock slop, Nayong bank slope, located in Liupanshui city, China, was selected as a case study and then was simulated in the light of the geological conditions for the study site. The main conclusions are presented as follows:

(1) Fracture network parameters for the study site were determined according to the data from geological survey. Based on the field investigation results, the overall stability of Nayong bank slope is mainly controlled by random fractures and bedding planes.

(2) The presented jointed rock slope model using an improved hybrid FEM, which is entirely different with the traditional model based on the framework of continuum mechanics. Thus the effects both of stochastic fractures and bedding planes were integrated into our numerical model. To analyze slope stability in a quantitative way, safety factor was calculated by the strength reduction method, and this procedure was also included in the numerical model.

(3) Numerical investigations were carried out to show the mechanical behaviors of jointed rock slope under different boundary conditions. Considering the loads induced by bridge foundation during construction period, slip surface mode in the case of loading condition is significantly different from that of in natural condition. Consequently, the impacts of bridge foundation on slope stability are reflected by the variation of slip surface patterns. It is reasonable to infer the bridge foundation would lead to a reduction of SF value.

(4) To compare the differences between intact and fractured slopes, we analyzed slope stability through different approaches. In the situation of intact rock slope, circular slip surface is the main instability mode, but for a fractured rock slope, polygonal slip surface is the major form. From the simulation results, we know that the value of SF calculated by hybrid FEM is smaller than that of by conventional FEM and LEM. This difference is mainly caused by not considering contributions of fractures.

References

1. L.R. Jing, Int..J. Rock. Mech. Min. Sci. **40**, 7 (2003)
2. P.M. Adler, J.F. Thovert, *Fractures and Fracture Networks* (Netherlands, Kluwer Academic Publishers, 1999)
3. J.C. Jaeger, N.G.W. Cook, R. Zimmerman, *Fundamentals of rock mechanics* (Hoboken, John Wiley & Sons, 2009)
4. B. Berkowitz, Adv. Water Resour. **25**, 8 (2002)
5. W.S. Dershowitz, H.H. Einstein, Rock Mech. Rock Eng. **21**, 1 (1988)

6. Q.H. Lei, J.P. Latham, C.F. Tsang, Comput. Geotech. **85** (2017)
7. T. Belytschko, W.K. Liu, B. Moran, (Hoboken, John Wiley & Sons, 2013)
8. R. Borst, Int. J. Numer. Methods Eng. **52**, 1 (2001)
9. D.V. Griffiths, P.A. Lane, Géotechnique **49**, 3 (1999)
10. E.M. Dawson, W.H. Roth, A. Dresche, Géotechnique **49**, 6 (1999)
11. S. Oberhollenzer, F. Tschuchnigg, H.F. Schweiger, J. Rock Mech. Geotech. Eng. **10**, 6 (2018)
12. D.Q. Li, D. Zheng, Z.J. Cao, X.S. Tang, K.K. Phoon, Eng. Geol. **203** (2016).
13. X. Fu, Q. Sheng, Y. Zhang, J. Chen, S. Zhang, Z. Zhang, Comput. Geotech. **92** (2017)
14. S. Metya, T. Mukhopadhyay, S. Adhikari, G. Bhattacharya, Comput. Geotech. **87** (2017)
15. T. Carlà, E. Intrier, P. Farina, N. Casagli, Int. J. Rock Mech. Min. Sci. **93** (2017)
16. J.M. Duncan, J. Geotech. Eng. **122**, 7 (1996)
17. A.W. Bishop, Géotechnique **5**, 1 (1955)
18. O.C. Zienkiewicz, C. Humpheson, R.W. Lewis, Géotechnique **25**, 4 (1975)
19. K. Zhang, P. Cao, J. Meng, K. Li, W. Fan, Rock Mech. Rock Eng. **48**, 2 (2015)
20. S. Wu S, L. Xiong, S. Zhang, Int. J. Geomech. **18**, 10 (2018)
21. O.C. Zienkiewicz, G.N. Pande, Int. J. Numer. Anal. Methods Geomech. **1**, 3 (1977)
22. C. Sun, J. Chai, Z. Xu, Y. Qin, X. Chen, Eng. Geol. **214** (2016)
23. M. Pirone, G. Urciuoli, Comput. Geotech. **102** (2018)
24. F. Tschuchnigg, H.F. Schweiger, S.W. Sloan, Comput. Geotech. **70**, 1 (2015)
25. F. Tschuchnigg, H.F. Schweiger, S.W. Sloan, Comput. Geotech. **70**, 2 (2015)
26. T. Ishida, M. Chigira, S. Hibino, Rock Mech. Rock Eng. **20**, 4 (1987)
27. M.A. Brideau, D. Stead, Rock Mech. Rock Eng. **43**, 3(2010)
28. G. Liu, X. Zhuang, Z. Cui, Eng. Geol. **225** (2017)
29. G. Yagoda-Biran, Y.H. Hatzor, Comput. Geotech. **71** (2016)
30. M. Barla, G. Piovano, G. Grasselli, Int. J. Geomech. **12**, 6 (2011)
31. E.G. Baghini, M.M. Toufigh, V. Toufigh, Comput. Geotech. **80** (2016)
32. X.P. Zhou, Y. Zhao, Q.H. Qian, Eng. Geol. **192** (2015)
33. P. Hamdi, D. Stead, D. Elmo, J. Rock Mech. Geotech. Eng. **7**, 6 (2015)
34. P. Fu, S.M. Johnson, C.R. Carrigan, Int. J. Numer. Anal. Methods Geomech. **37**, 14 (2013)
35. L. Guo, X. Hu, L. Wu, X. Li, H. Ma, Int. J. Geomech. **18**, 10 (2018)
36. T. Miyoshi, D. Elmo, S. Rogers, J. Rock Mech. Geotech. Eng. **10**, 6 (2018)
37. . Elmouttie, G. Krähenbühl, A. Soliman, Comput. Geotech. **76** (2016)
38. J.C.S. Long, J.S. Remer, C.R. Wilson, P.A. Witherspoon, Water Resour. Res. **18** (1982)
39. K. Bisdom, H.M. Nick, G. Bertotti, Comput. Geosci. **103** (2017)
40. L.Y. Wang, W.Z. Chen, X.Y. Tan, X.J. Tan, J.P. Yang, X. Zhang, Bull. Eng. Geol. Environ. **5** (2019)
41. J.C. Brewer, P.A. Lagace, J. Compos. Mater. **22**, 12 (1988)
42. S.J. Feng, Z.W. Chen, H.X Chen, Q.T. Zheng, R. Liu, Eng. Geol. **241** (2018)

Seismic analysis on cross joint of utility tunnel

Chang-xiang Wang[1,*], *Meijing* Wang[2] , and *Jianwen* Liang[2]

[1]North China Municipal Engineering Design & Research Institute Co., Ltd., Tianjin 300074, China
[2]School of Civil Engineering, Tianjin University, Tianjin 300350, China

Abstract. The finite element software ABAQUS is used to establish the seismic time-history analysis model of the utility tunnel joints. The viscoelastic boundary is employed to input the seismic wave through the equivalent nodal force. Considering the nonlinearity of the soil and concrete materials, the internal force of the cross-shaped utility tunnel is calculated, and the influence effect and influence range of the cross joints of the utility tunnel are also studied. Studies have shown that the internal forces near the cross joints increase significantly. The research has certain reference value for the seismic design of the cross joints in utility tunnel.

1 Introduction

Over the years, with the rapid development of utility tunnel construction, the seismic performance of the cross joints of utility tunnel has received extensive attention. Earthquakes occur frequently in China, the seismic analysis of utility tunnels and their cross-joints is of great significance for the seismic design of utility tunnel. Takada [1] studied the seismic design principles and methods of underground pipelines, and proposed initial practical parameters and formulas, and compared the seismic design of the lifeline in terms of earthquake ground motion input, stratum and pipeline deformation from the perspective of application. Yue [2] and Yue et al [3] used the comparative experiment and three-dimensional finite element model to achieve the research and analysis on the utility tunnel, and considered many factors affecting the structural response in the analysis, such as the influence of artificial boundary conditions, contact surfaces and soil plasticity. Through theoretical derivation, different ground motion input methods of underground structure are discussed and the displacement input is recommended as a more accurate input method. Liu et al. [4] studied the error source of the quasi-static method of underground structures through finite element software comparison analysis, and improved and optimized the seismic wave loading method of the reaction acceleration method. Shi et al. [5] established a three-dimensional dynamic finite element numerical model for a dual-compartment utility tunnel and applied a pulse load to the boundary of the model to generate Rayleigh waves, and compared that with the conventional time-history analysis considering only the bottom lateral action, Through this comparison, the effects of different soil constitutive models, buried depths, incident angles and different cross-sections on the seismic behavior of the utility tunnel under the combined action of Rayleigh wave and

* Corresponding author: wangcx87@163.com

bottom seismic acceleration are studied. However, there are relatively few related researches on complex cross-joints of utility tunnel. In this paper, the finite element software ABAQUS is used to establish a nonlinear time-history analysis model for the cross joints in utility tunnels. Considering the nonlinearity of the soil and concrete materials, the dynamic analysis on joints in utility tunnel is carried out by using the viscoelastic boundary and the method of inputting seismic waves by equivalent nodal force. Based on that, the influence effect and influence range of the cross joints of utility tunnels are studied. This study has certain reference value for the seismic resistance of the cross joints of utility tunnels.

2 CALCULATION METHOD

2.1 Viscoelastic artificial boundary

In this paper, the nonlinear time-history analysis method is used and a three-dimensional soil-pipeline model using the finite element software ABAQUS is established. By applying a viscoelastic boundary and equivalent nodal force [6] at the boundary of the soil layer, equivalent linearization analysis [7] of soils is performed.

2.2 Seismic input of viscoelastic artificial boundary

The seismic wave acting at the viscoelastic artificial boundary consists of two parts, one part is the original incident seismic wave, and the other part is the reflected wave generated by the internal seismic wave at the boundary position, and the equivalent node force applied is to simulate the effects of these two waves, thus to ensure that the displacement and stress at the artificial boundary are consistent with the original free field. In this paper, the response of the free field is first obtained according to the EERA program [8], and then converted into the equivalent nodal force applied at the viscoelastic boundary node to perform the input of the seismic wave.

2.3 Damage plasticity model for concrete in ABAQUS

The concrete damage is based on the concrete damage plasticity CDP model [9-10] in ABAQUS software. The CDP model in ABAQUS combines isotropic tensile plasticity theory, isotropic compression plasticity theory and isotropic damage theory to describe the inelastic behavior of concrete. It is a continuum model considering plasticity and damage [11], which is mainly used to describe unidirectional loading, repeated loading and dynamic loading. In use, it is necessary to input the corresponding relationship between yield stress and inelastic strain, damage factor as well as stiffness recovery factor in ABAQUS. ABAQUS automatically converts inelastic strain into equivalent plasticity strain when performing analysis, and determines the elastic modulus of concrete according to the stiffness recovery factor and the damage factor. Relevant parameters are determined according to the relevant specification [12].

3 NONLINEAR SEISMIC RESPONSE ANALYSIS OF CROSS-JOINT UTILITY TUNNEL

3.1 Calculation parameters of utility tunnel cross-joint model

The one-way utility tunnel in the X-axial direction is named as a one-way X-direction utility tunnel; the one-way utility tunnel in the Y-axial direction is named as a one-way Y-direction utility tunnel; as shown in Fig. 1, in cross-shaped utility tunnel, the utility tunnel in X direction is named as the cross X-direction utility tunnel, and that in the other direction is named as the cross Y-direction utility tunnel.

Parameters of the horizontal layered site are shown in Table 1. The utility tunnel has a buried depth of 3 m, a cross section of 6 m×6 m and a thickness of 0.3 m. The concrete material parameters are shown in Table 2. The seismic wave uses an El Centro wave with a peak value of 0.1 g, which is incident perpendicularly from the bedrock surface and vibrates along the X axis. The acceleration time history curve is shown in Fig. 2. The equivalent linear parameters of soil materials are shown in Fig. 3.

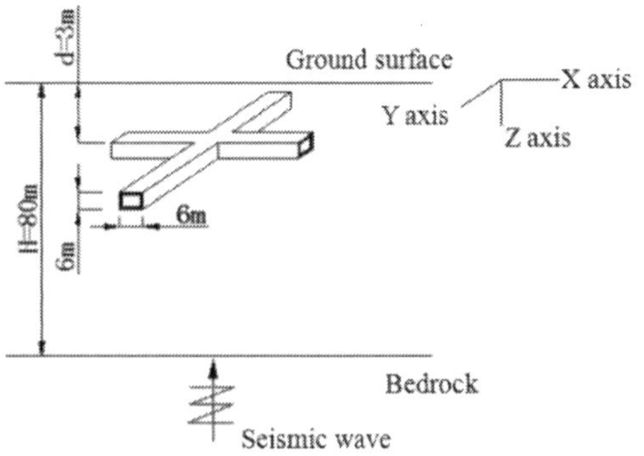

Fig. 1. Cross-joint utility tunnel problem model.

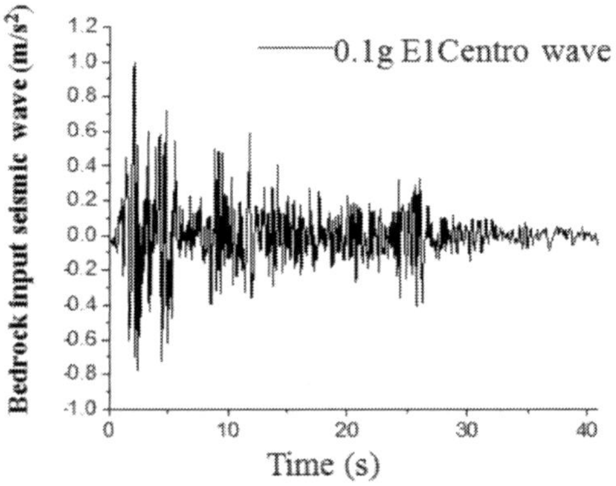

Fig. 2. 0.1 g El Centro wave.

Table 1. Soil layer material parameters.

Soil layer number	Category	Thickness (m)	Shear wave velocity(m/s)	Density (kg/m3)	Damping ratio
1 (Ground surface)	I	10	150	1750	0.05
2	I	10	175	1775	0.05
3	I	10	200	1800	0.05
4	I	10	250	1850	0.05
5	II	10	300	1900	0.05
6	II	10	350	1950	0.05
7	II	10	400	2000	0.05
8	II	10	450	2050	0.05
9（Bedrock）		∞	500	2100	0.02

Table 2. Concrete material parameters

Concrete	Density（kg/m3）	Elastic Modulus（MPa）	Damping Ratio
C60	2600	35500	0.05

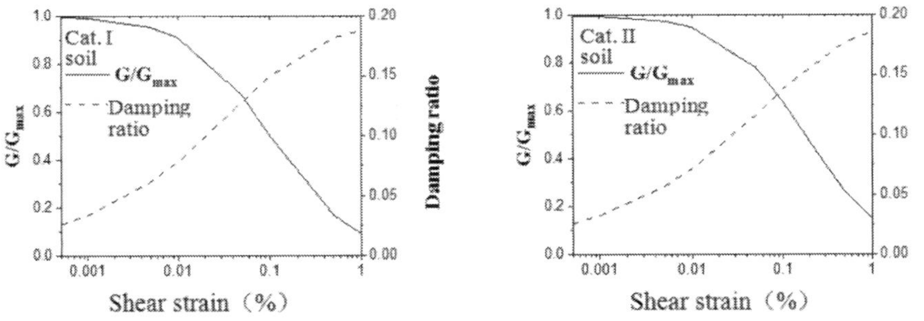

Fig. 3. Soil nonlinear characteristic curve.

3.2 Model establishment

According to the above analysis, the size of the model is 170 m×90 m×80 m (X×Y×Z), the one-way X-direction utility tunnel is 170 m long and the one-way Y-direction utility tunnel is 90 m long, and the cross X-direction utility tunnel and cross Y-direction utility tunnel are respectively 170 m and 90 m long. The utility tunnel uses the S8R shell unit. The utility tunnel meshing is shown in Fig. 4, and the soil meshing is shown in Fig. 5.

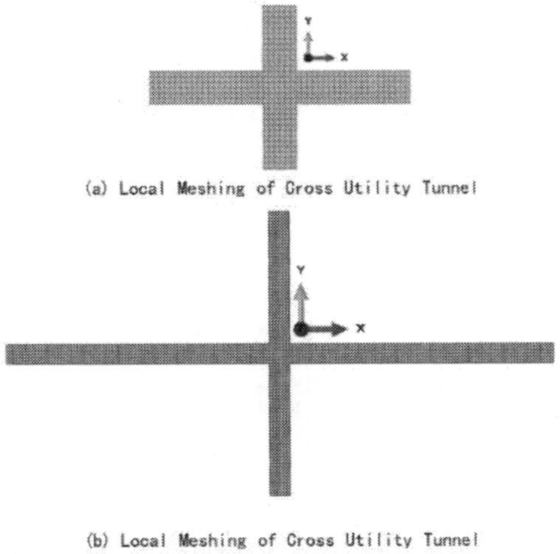

Fig. 4. Utility tunnel meshing.

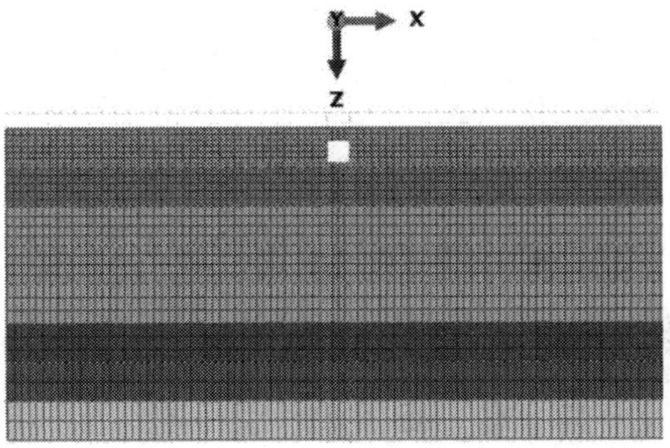

Fig. 5. Meshing.

3.3 Internal force analysis of X-direction utility tunnel at the cross joint of utility tunnel

The axial force *N*, the shear force *V* and the bending moment *M* of the cross one-way X-direction utility tunnel when maximum seismic motion emerges in X-direction are compared with the internal force of the one-way X-direction utility tunnel to explore the influence range and influence effect of the utility tunnel cross-joint on the cross X-direction utility tunnel. Analytical section positions of cross X-direction utility tunnel are as shown in Fig. 6.

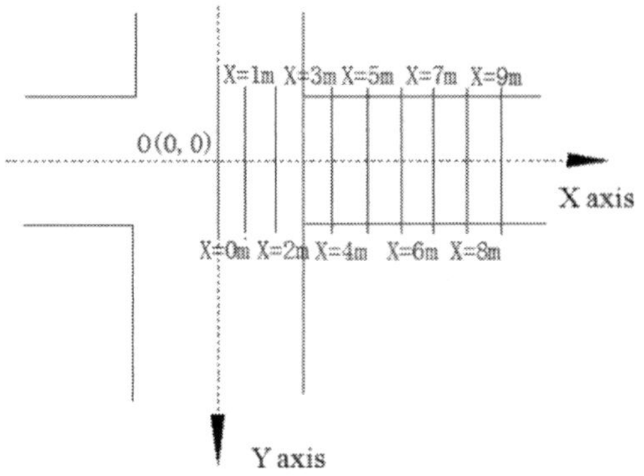

Fig.6. Schematic analytical section of cross X-direction utility tunnel.

As shown in Fig. 7, the existence of the cross-joint increases the axial force *N* of the cross X-direction utility tunnel. At the cross joint (X=3 m), the increment of axial force *N* reaches the maximum increased by about 280%; When the distance from the cross joint increases, the influence to the axial force is weakened. At a distance of 35 m from the axial center section of the utility tunnel, the axial force N increases by about 30%, and at 60 m, it increases by 16%.

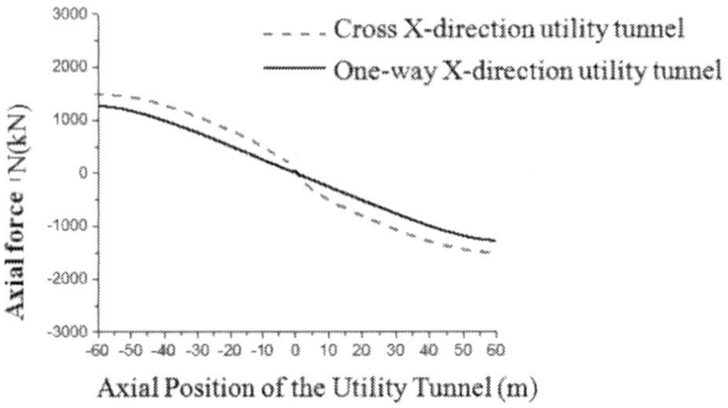

Fig. 7. Axial force comparison of X-direction utility tunnel

As presented in Fig. 8, the shear force *V* of the cross X-direction utility tunnel in the joint area declines greatly, close to 0; at the cross-joint position (X=3 m), the shear force V increases sharply, which is 1500 kN, increased by about 50% relative to the one-way X-direction utility tunnel; then the shear force *V* is gradually reduced, and the difference between it and that of the one-way X-direction utility tunnel is within 5% at a distance of 35 m from the axial center section of the utility tunnel.

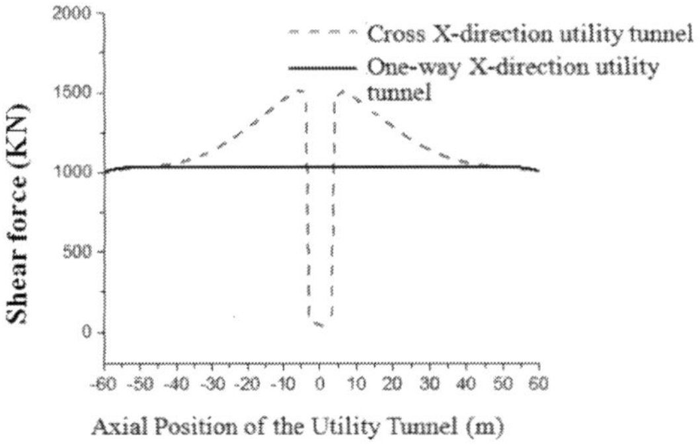

Fig. 8. Shear force *V* Comparison of X-direction utility tunnel.

As given in Fig. 9, with respect to the axial force *N* and the shear force *V*, the existence of the cross joint makes the increase effect of the bending moment M of the cross X-direction utility tunnel most obvious, and becomes the control internal force of the section; at the cross-joint position (X=3 m), the bending moment *M* is 9700 kN·m, and then the increasing effect of the joint is rapidly reduced. At a distance of 35 m from the axial center section of the utility tunnel, the difference between the bending moment *M* and that of the one-way X-direction utility tunnel is within 5%.

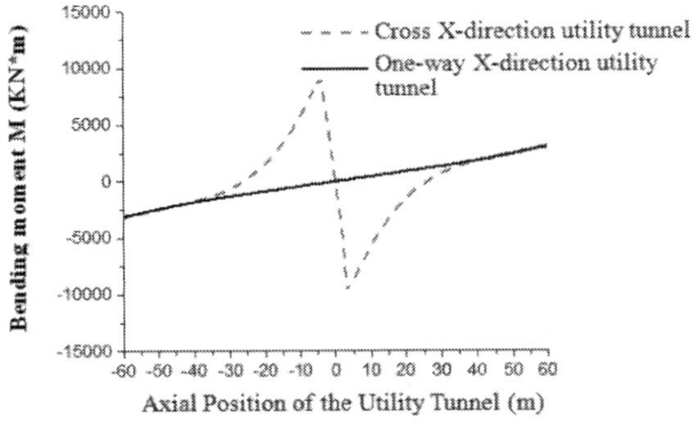

Fig. 9. Bending moment *M* comparison of X-direction utility tunnel.

In summary, the influence range of the cross joint is 35×2=70 m. The cross joint increases the axial force *N* and the bending moment *M* of cross X-direction utility tunnel, and decreases the shear force *V* inside the cross-joint area while increases that outside the area. And the internal force increase effect is most obvious at the cross-joint position.

Fig. 10 shows that the shear force FV(X) of the cross X-direction utility tunnel increases inversely, which is -52.6 kN at the center position (X=0m), and increases by about 150% in the reverse direction; The increase effect at the cross-joint position (X=3m) is most pronounced, with a shear force FV(X) of -93 kN i.e. a reverse increase by approximately 350%. At a distance of 7m from the axial center section of utility tunnel, the maximum positive increase emerges, that is 45.8 kN; then the increase effect is gradually weakened, the analysis unit shear force FV(X) difference between the cross X-direction utility tunnel and one-way X-direction utility tunnel is within 5% at a distance of 40 m from the axial center section of the utility tunnel.

It is presented in Fig. 11 that the torque FT change of the cross X-direction utility tunnel is similar to the shear force FV(X), and the torque FT in the cross-joint area increases inversely, and is -9.1 kN at the center position (X=0 m), a reverse increase of about 180%; the increase effect at the cross-joint position (X=3 m) is most obvious, with a torque FT of -11.8 kN·m and a reverse increase of about 260%. At a distance of 7m from the axial center section of utility tunnel, the maximum positive increase emerges, that is 6.82 kN·m; then the increase effect is gradually weakened, the analysis unit Torque FT difference between the cross X-direction utility tunnel and the one-way X-direction utility tunnel is within 5% at a distance of 40 m from the axial center section of the utility tunnel.

To conclude, the cross joint exerts influence on the shear force FV(X) and torque FT in a larger area, which is 40×2=80m. The cross joint increases the shear force FV(X) and the torque FT of the analysis unit, wherein reverse increase appears in the cross-joint area and the increase effect at the cross-joint position is the largest.

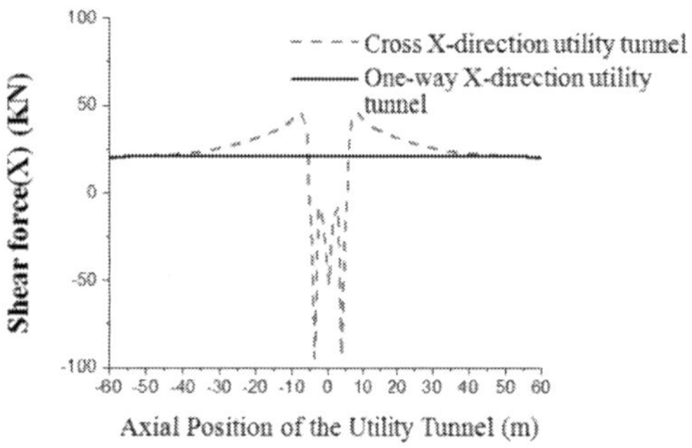

Fig. 10. Analysis unit shear force FV (X) comparison of X-direction utility tunne.

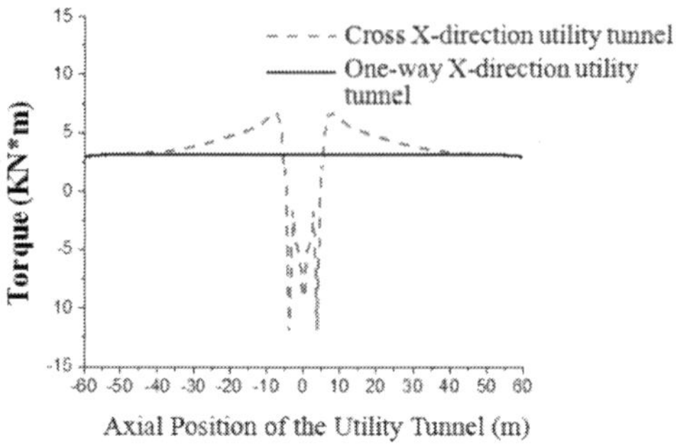

Fig. 11. Analysis unit torque FT comparison of X-direction utility tunnel.

3.4 Internal force analysis of Cross Y-direction utility tunnel

Fig. 12 shows the analysis section position of the cross Y-direction utility tunnel.

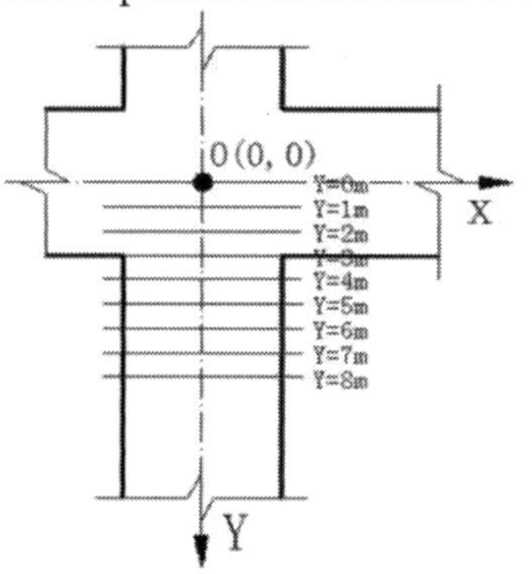

Fig. 12. Schematic analytical section of cross Y-direction utility tunnel.

As shown in Fig. 13, the existence of the cross joint increases the axial force FN in the cross-joint area. At the cross-joint position (Y=3m) and the axial force FN is the largest, which is -240.6 kN, increased by about 750% compared with the one-way Y-direction utility tunnel; The axial force FN outside the cross-joint area is smaller than that of the one-way Y-direction utility. With the distance from the cross joint increases, the cross-joint effect on the axial force FN decreases gradually. At Y=35 m, the axial force difference of analysis unit at the corresponding position of cross Y-direction utility tunnel and one-way Y-direction utility tunnel is within 5%.

As presented in Fig. 14, the existence of the cross joint makes the shear forces FV of the cross Y-direction utility tunnel smaller than the one-way Y-direction utility tunnel, but the shear force FV at the cross-joint position (Y=3 m) is larger than those at other positions of the analysis unit in the cross-joint area due to stress concentration. At Y=35 m, the analysis unit shear force difference between the cross Y-direction utility tunnel and the one-way utility tunnel is within 10%.

Fig. 15 shows that the influence of the cross joint on the bending moment FM of the analysis unit is similar to that on the axial force FN. The maximum bending moment FM is also found at Y=3 m, which is -41.2 kN·m, increased by about 136 % relative to the one-way Y-direction utility tunnel; except locations near the cross-joint position , the bending moments FM at other locations are smaller than those of the one-way Y-direction utility tunnel. At Y=30 m, the bending moment difference with the one-way Y-direction utility tunnel is within 5%.

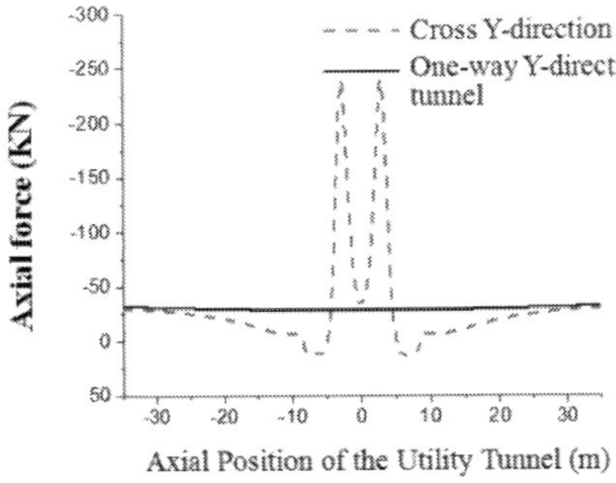

Fig. 13. Analysis Unit Axial Force FN Comparison of Y-direction Utility Tunnel

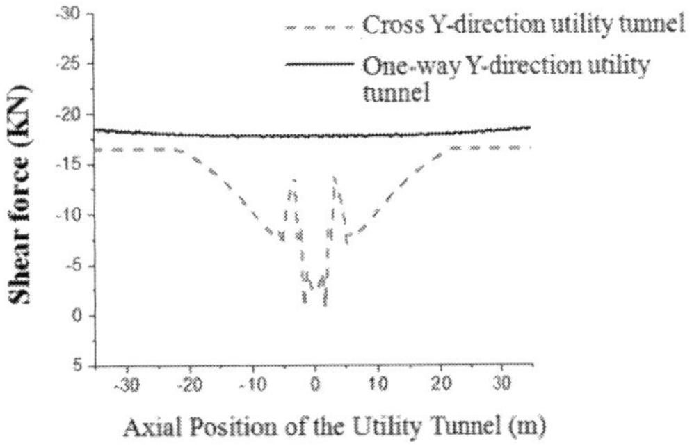

Fig. 14. Analysis Unit Shear Force FV Comparison of Y-direction Utility Tunnel

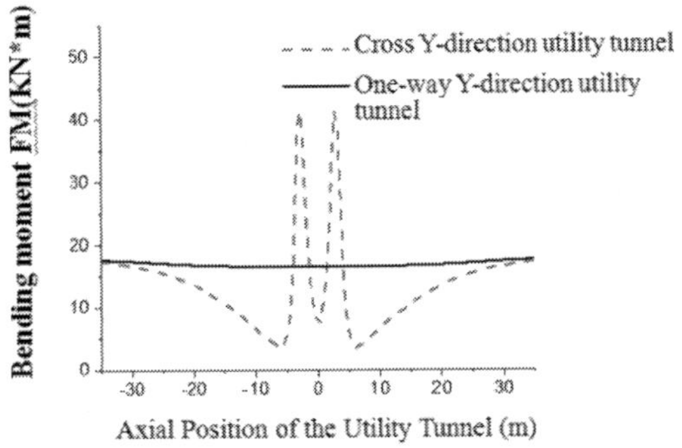

Fig. 15. Analysis Unit Bending Moment FM Comparison of Y-direction Utility Tunnel

To sum up, the existence of the cross joint reduces the axial force FN, the shear force FV and the bending moment FM of the cross Y-direction utility tunnel except locations near the cross-joint position. This is because a larger stiffness at the cross-joint position makes the shear deformation smaller than the one-way Y-direction utility tunnel, thus leads to the decrease of the axial force FN, shear force FV, and bending moment FM outside the cross-joint area, while an increase of the axial force FN, shear force FV, and bending moment FM appears due to stress concentration at the cross-joint position. The influence range of the cross joint on the shear force FV is larger than that on the bending moment FM and the axial force FN, but compared with the influence effect of the shear force FV at the cross-joint position (Y=3 m), the influence effect of 10% shear force difference at Y=35 m can be neglected. It can be considered that the influence range of the cross joint on the cross Y-direction utility tunnel is 35×2=70 m.

As shown in Fig. 16 and Fig. 17, in addition to the influence on the axial force FN, the shear force FV and the bending moment FM on the cross section of the cross Y-direction utility tunnel, the existence of cross joint also generates section shear force V and the torque T. For the shear force V, the maximum value appears at the cross-joint position (Y=3m), which is 194.7 kN, and then it decreases gradually. For the torque T, the maximum value also appears at Y=3 m, which is 11360 kN·m, and it decreases with the increase of the distance from the cross joint. In the range of Y=20 m, both shear force V and torque T decline more steeply, which is the main influence area of the cross joint; in contrast to the shear force V and the torque T at the cross-joint position, the shear force V and torque T generated at Y=35 m is negligible. The torque T generated by the cross joint becomes the control internal force of the section, and its increase effect is far greater than the reduction effect of the cross joint on the axial force FN, the shear force FV and the bending moment FM of the cross section, so the cross joint also has an adverse effect on the cross Y-direction utility tunnel.

The calculation shows that the axial force FN in the cross section of the cross Y-direction utility tunnel increases in the cross-joint area (between Y=-3 m and Y= 3 m), and the maximum value of 466.1 kN emerges at the cross joint (Y=3 m); axial force FN outside the cross-joint area is smaller than that of the one-way Y-direction utility tunnel, and with the increase of the distance from the cross joint, the reduction effect of the cross joint on the axial force FN is gradually weakened; the maximum axial force FN of each analysis section appears near the lower corner.

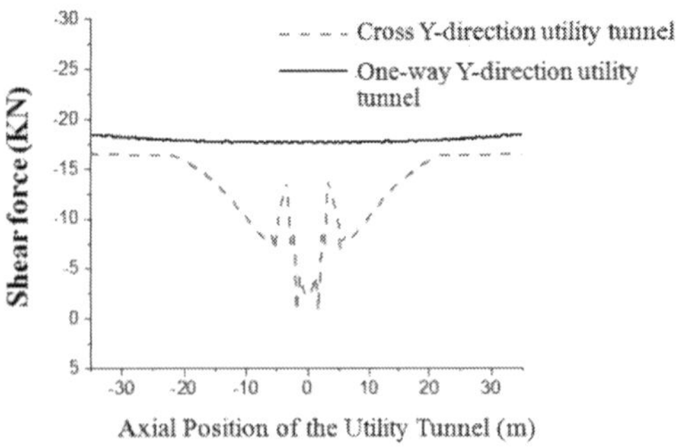

Fig. 16. Shear force *V* distribution of Cross Y-direction utility tunnel

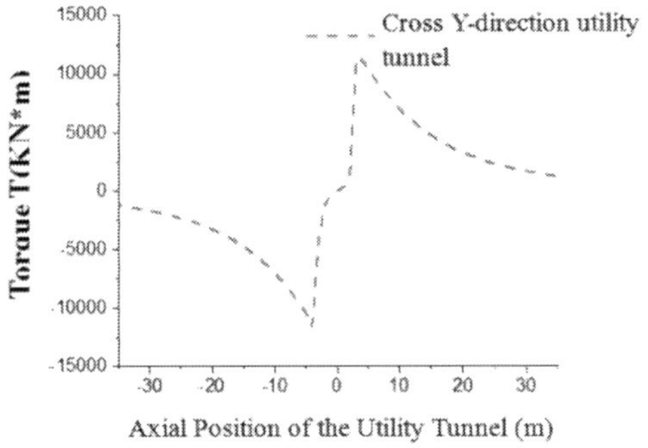

Fig. 17. Torque *T* distribution of Cross Y-direction utility yunnel

4 Conclusions

The research shows that the cross joint increases the internal force of the utility tunnel which vibrates in the axial direction of the utility tunnel, and the bending moment *M* becomes the control internal force of the section. The increase effect on internal force is the most obvious at the cross-joint position. The presence of the cross joint causes the shear force *V* and the torque, wherein the torque becomes the control internal force of the section in another utility tunnel.

References

1. S. Takada. J. Tunnel Trans. **1991** 45 (1991)
2. Q. X. Yue. Thesis. Shanghai: Tongji University (2007)
3. Q. X. Yue, J. Li. J. Tongji Uni (Natural Science Edition) **7** 285 (2009)
4. R. S. Liu, S. Q. Hu, H. B. Shi. Chinese J. Geotech. Eng. **29** 237(2007)
5. Y. Z. Shi, A. Q. Sun, S. Z. Lin , et al. World Earthq. Eng. **33** 196 (2017)
6. J. B. Liu, Y. X. Du, Q. S. Yan. J. Disas. Prev. Mitig. Eng. **27** 37 (2007)
7. J. B. Liu, Y. Gu, Y. X. Du. Chinese J. Geotech. Eng. **28** 1070 (2006)
8. J. P. Bardet, K. Ichii, C. H. Lin. *EERA-A computer program for equivalent-linear earthquake site response analyses of layered soils deposits* (2000)
9. J. Lubliner. J. Oliver. S. Oller. E. Oñate. Int. J. Solids and Structs **25** 229 (1989)
10. J. Lee, G. L. Fenves. J. Eng. Mech. ASCE **124** 892 (1998)
11. W. Liu, M. Xu, Z. F. Chen. Industrial construction **44** 167 (2014)
12. Code for design of concrete structures GB 50010-2010

Effect of the geotextiles reinforcement area on the stabilization of soil deformations in Chlef city (Algeria)

Zohra Boutaraa[1**], *Ahmed Arab*[1] *and Mohammed Chemmam*[1]

1 Laboratoire des Science des Matériaux et de l'Environnement, Université HBB de Chlef, 02000 Chlef, Algérie

Abstract. This communication presents a laboratory study on the effect of the geotextiles reinforcement area on the improvement of the mechanical proprieties of a sandy soil, especially on the stabilization of soil deformations that affect road pavement in Chlef city (Algeria), which is a prone area to seismic risk. The study is carried out via a monotonic undrained tests using the triaxial apparatus. Reconstructed soil samples having a relative density corresponding to two soil states: loose (Dr = 20%) and medium dense (Dr = 50%) are tested. The aim of the study is to highlight the improvement that can bring the increase of the area of geotextile reinforcement on the behavior of the studied soil. Results show that the reinforcement area has a great positive effect on the increase of the soil resistance, traduced by the stress deviator and by the friction angle, and on the decrease of the pore pressure that reduce the liquefaction risk.

1. Introduction

Soil reinforcement technics represent an efficient tool that can be used in the aim to resolve instability problems encountered in low mechanical quality soils, especially in prone area to seismic risk. These technics that have been developed considerably all over the world are actually used as an incorporate part of civil engineering projects since it can increase the soil mechanical properties and give it a better behavior, particularly in seismic zones, where there are used as a seismic risk management tool. Among soil reinforcement process frequently used we are interested by geotextiles, which consists in introducing in the soil one or more layer number of geotextiles. This process gives very good results in superficial layers of sandy soils, where the confinement has low values [1]. Several researchers have studied the reinforcement of Chlef sandy soil using geotextiles. However their interest pointed on the effect of the increase of geotextiles layers number on the soil mechanical improvement [2, 3]. In Chlef city, the road pavement of the main express national road (RN4) that longe Chllif river is instable and important deformations are constantly emerging (Fig.1). These deformations are due to the intensive dynamic load supported by the pavement and to the low soil quality which is an alluvial quaternary deposit with low compactness, mainly formed of sand, gravel and clay and which superficial layers have low shear wave propagation (Vs = 230~ 350 m/s), and to the high settlement potential in this section [4]. Note that Chlef city is located in a high seismic zone [5] and since the soil saturation degree is sufficiently high in these layers adjacent to Chellif river, the liquefaction risk is also important in case of a seismic event. Therefore, in this work our interest is oriented on the effect of the increase of the geotextiles reinforcement area on the behavior of this soil and we propose to study the effect of the geotextiles reinforcement area on the undrained behavior of Chlef sandy soil under monotonic loading.

Figure 1: Important deformations of a section of RN4 express road in Chlef city (Algeria)

* Corresponding author : z.boutaraa@univ-chlef.dz

2. Experimental Procedure and materials

A laboratory study is carried out using the triaxial apparatus in undrained monotonic conditions. Soil samples, constituted by Chlef sand (γ =2.69 g/cm^3) and 20% of silt (γ =2.70 g/cm^3,) are reconstituted using under-compaction method [6] and saturated before tested. The saturation degree affecting significantly the soil liquefaction resistance is evaluated by measuring the Skempton's coefficient B after consolidation [7]. Two relative densities (Dr=20 and 50 %) corresponding to two soil states (loose and medium dense) are experimented.
The reinforcement is made using a geotextile type NOTEX GX 100/50 which characteristics are given in Table 1. One geotextile layer is placed at the sample middle high (Fig.2).Three samples are tested for every soil state: a first sample without reinforcement (SR), a second one in which the reinforcement géotextile area is equal to 2/3 of the sample section (2/3 S) and a last one in which the reinforcement géotextile area is equal to the totality of the sample section (S). Note that these géotextile areas are chosen because of the easiness of their cutting in the laboratory and positioning in the sample.

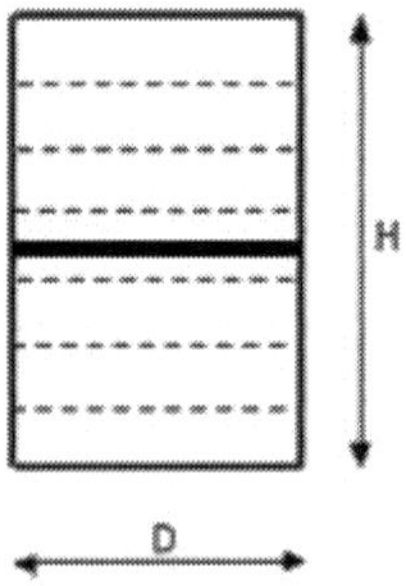

Figure 2: Position of the geotextiles in the soil samlpe

Table 1: Technical characteristics of the used geotextile

Characteristics	Norm	Value	Tolerance
Mass per unit area (g/m2)	Standard ISO 9864	300	± 10%
Thickness under 2 kPa (mm)	Standard 964-1	1,25	± 20%
Surface porosity (%)		43	
Tensile rupture strength (kN/m)	Standard EN ISO10319	SP : 100	Min. Val
		ST : 50	Min. Val
Tensile strength at 2 % strain SP (kN)		16	
Tensile strength at 3 % strain SP (kN)		23	
Tensile strength at 5 % strain SP (kN)			Val. Min.
Strain at failure in tension (%)	Standard EN ISO10319	38	± 20%
			± 20%
Dynamic perforation (mm)	Standard EN ISO13433	19	± 20%
Static puncture CBR (KN)	Standard EN ISO12236	2,5	-10 %

3. Study results

3.1. Effect of the geotextile reinforcement area on the soil resistance

Figure 3a shows the evolution of the stress deviator q in the loose soil samples before and after its reinforcement. It is concluded that the reinforcement area affect considerably the soil resistance since the maximum value of q reaches 94 kPa in the no reinforced sample, then rises to the values of 120 and 186 kPa in the reinforced samples on an area of 2/3 and of the total sample section, respectively. Figure 3b shows the pore pressure U evolution in the no reinforced sample reaching maximal values of 190 corresponding to an axial deformation ε_l of 5%, then U decreases linearly until 140 kPa. U maximal value in the samples which are reinforced on an area equal to 2/3 and to the totality of the sample section are less important: 85 and 80 kPa, respectively. In this latter sample, the maximum axial deformation ε_l is equal to 3%, and the decrease of U is more pronounced until its annulation.

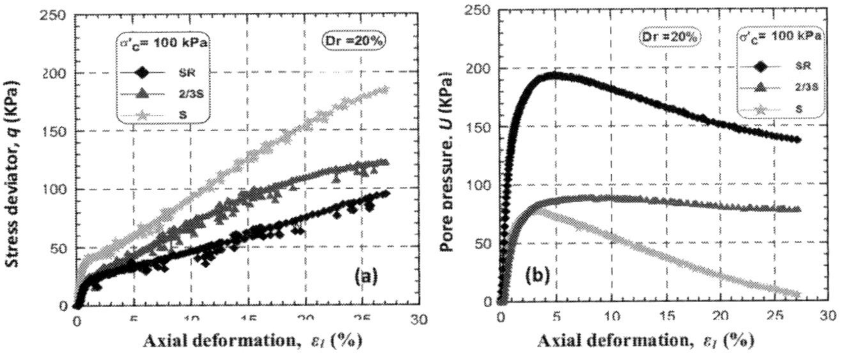

Figure3: Reinforcement area effect on the undrained behavior of the loose soil sample: (a) Evolution of the stress deviator and (b) Evolution of the pore pressure

In the case of the medium dense samples, the increase of the geotextile reinforcement area effect becomes more pronounced. Indeed, the maximum value of q that do not exceed 150 kPa in the no reinforced sample, rises the values of 180 and 200 kPa in the samples reinforced on the 2/3 and on the totality of their sections, respectively (Fig. 4a). It can be observed that the second sample has initially a dilatant behavior and has a higher resistance than the last one (reinforced on its total section) until a ε_l value of 15%. After this a different behavior is observed; an increase of the sample resistance versus of the reinforcement area, which is due to the geotextile role that deprives radial deformations in the sample and consequently the soil has contracting behavior. Concerning the pore pressure evolution, it can be observed that the reinforcement area has an important effect on its decrease. Indeed, for about a value of 2.5% of ε_l, U reaches maximal values of 79, 75 and 72 kPa in the no reinforced sample, the reinforced on an area of 2/3 and on the totality of the sample section, respectively (Fig. 4b).

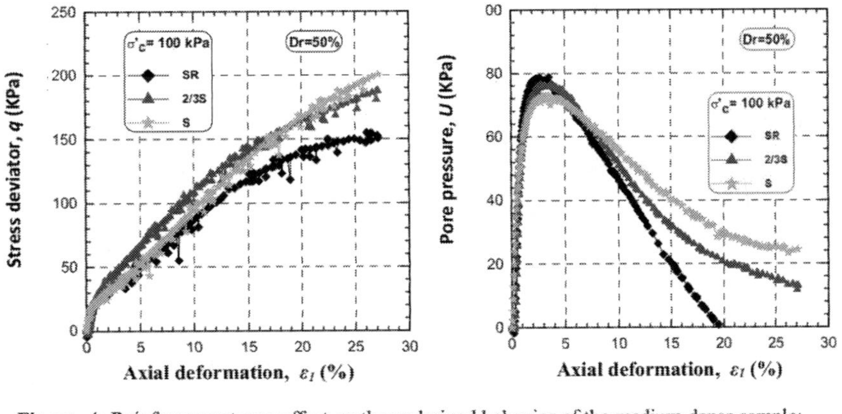

Figure 4: Reinforcement area effect on the undrained behavior of the medium dense sample: (a) Evolution of the stress deviator and (b) Evolution of the pore pressure

3.2. Effect of the geotextile reinforcement area on the soil friction angle

The introduction of geotextile reinforcement in the sample conducts to a modification of the soil mechanical characteristics. Among these characteristics we are interested by the friction angle φ (eqt 1) because it is the most important one that affect the behaviour of sandy soils as these encountered in the studied area. Other parameters can be measured and highlighted as the secant modulus for instance, but due to the importance of the friction angle and the deviator stress that characterize the resistance of the studied soil, we discussed just these two parameters in this paper.

$$\varphi = \mathbf{Arcsin\ (3^* \eta /6+ \eta)} \qquad (1)$$

Where η is given by equation 2:

$$\eta = q / \sigma'eff \qquad (2)$$

Where q is the deviator stress and $\sigma'eff$ is the effective stress.

As an example in the medium dense sample, this evolution is exponential. Since φ value is equal to 26.9° in the no reinforced sample then it becomes equal to 28.47° and to 29.26° in the samples reinforced on the 2/3 and on the totality of their sections, respectively (Fig.5). This result confirms the positive effect of the increase of the geotextile reinforcement area on the mechanical behavior of the studied soil.

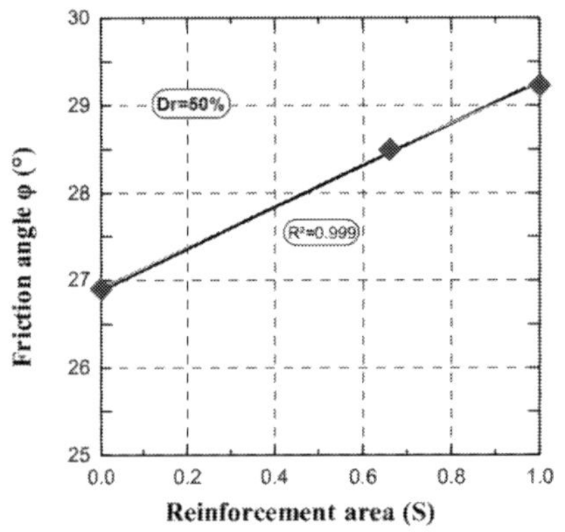

Figure 5: Reinforcement area effect on the soil friction angle

4. Conclusions

This communication gives the results of a laboratory study on the effect of the geotextiles reinforcement area on the improvement of the mechanical proprieties of Chlef sandy soil. The soil samples are tested under undrained monotonic conditions using the triaxial apparatus and for two relative densities corresponding to a loose and medium dense soil states. The used reinforcement areas are equal to 2/3 and to the totality of the sample section. The obtained results show that the increase of the geotextiles reinforcement area has a positive effect on the improvement of the mechanical properties of this soil after reinforcement. This improvement affects the soil resistance traduced by the stress deviator q and the soil friction angle φ. In another hand, the decrease of the pore pressure U indicates the diminution of the liquefaction risk that can occurs during a seismic event in Chlef city (Algeria) which is a prone area to seismic risk. The solution proposed in this communication to resolve soil deformations affecting the main express national road (RN4) in Chlef city, can easly be applied on real scale using geotextiles layer that will be placed under the road pavement on a depth of about 0.70 m on all the area affected by vertical or horizontal deformations.

References

1. Arab A., Belkhatir M., Sadek M. Geot. Geol. Eng. **34**, 347-358 (2016).
2. Benessalah I., Arab A., Villard P. et al. Geot. Geol. Eng. **34**, 6: 1775-1790 (2016)
3. Brahim AEK., Arab A., Belkhatir M., Shahrour I. J. Earth Sci. **27**, 6: 1060-1070 (2015)
4. Woodward-Clyde Consultants. Microzonation Sismique d'Echllif - Algérie. Alger, p 166 (1984)
5. RPA. Règles Parasismiques Algériennes. Centre National de Recherche Appliquée en Génie Parasismique, Alger (2003)
6. Ladd R. S. Geot. Tes. J, ASTM **1**(1): 16-23 (1978).
7. Skempton AW. Géot. Vol.**4**,4 : 143-147 (1954)

Disturbance effect analysis of pipe jacking construction based on failure approach index

Geng Liu[1], *Zhengchun* Yang[2], *Huayan* Yao[1,*], *Zhenhua* Zhang[1], *Jingfeng* Wang[1]

[1] Shool of Civil Engineering, Hefei University of Technology, 23009, Hefei, China
[2] China Water Huaihe Planning, Design and Research Co., Ltd, 230601, Hefei, China

Abstract. In order to evaluate the influence of pipe jacking construction process on ground subsidence and surrounding buildings, the numerical simulation of the construction process was carried out by FLAC^{3D}, and the failure approach index (FAI) of the rock and soil around the pipeline was calculated to describe quantitatively their risk and damage state. The numerical results show that the final settlements of the buildings around the pipeline are less than the buildings' safety control values. The rock or soil mass within 0.8 times of the pipe diameter are in the excavation damage zone and disturbance zone (0.8≤FAI≤1.3), while outside 0.8 times of the pipe diameter are basically undisturbed (FAI≤0.8).

1 Introduction

Pipe jacking has become popular in the USA during 1950s[1]. As a trenchless technology, pipe jacking is widely used in modern municipal pipe networks. Pipe jacking, compared with open cut method, significantly reduces the destruction of city buildings and roads caused by underground pipeline projects, as well as the impact on urban traffic [2]. However, there is inevitable disturbance to stratum structure, especially when the environment is very complex, inhomogeneous ground settlement and uneven deformation and crack of ground buildings in the course of pipe jacking construction. Many scholars have done plenty of researches on the basic rules of the stratum structure disturbance and ground settlement caused by pipe jacking constructions[3~12], but the complex environment and specific engineering requirements are still needed to be considered in detail. Using numerical simulation method, construction process in a complex environment can be simulated according to engineering practice. In the simulation results, the yield states of rock and soil mass can be revealed through plastic zone. But the dangerous level and damage degree cannot be reflected accurately by plastic zone. In order to describe it more intuitively and accurately, Zhang et al.[13~15] put forward a definition called failure approach index (FAI) and deduced its calculation formulas with several examples to verify the rationality and applicability. Based on FAI, Yao et al.[16] analyzed the stability of soils around shield tunnel during excavation. FAI has the ability to describe quantitatively the dangerous level or

* Corresponding author: yaohuayan@hfut.edu.cn

damage degree of soils or rock remaining in the elastic stage before or after its stress state reaches plastic state. Therefore, FAI is capable of providing scientific basis to the evaluation of stability of soil or rock impacted by geotechnical engineering construction.

In this paper, for analyzing the influence of pipe jacking on surroundings more accurately, FLAC3D is used to simulate a pipe jacking construction process. The damage range and degree of the soil around the pipeline is quantified by the FAI index. In addition, combining with the site monitoring data, the influence of pipe jacking on surrounding constructions are also revealed clearly.

2 Project profile

The pipe jacking construction technology is applied to the transformation of the rain and sewage pipeline nearby a subway station in Hefei City. The construction environment is complex with many buildings adjacent to the new pipeline. It could be a threat to the safety of nearby buildings and roads if pipe jacking construction induces uneven deformation of ground and foundation.

A rain pipeline of diameter 2.5m and a sewage pipeline of diameter 0.8m are laid in parallel in the south of a subway station. Two segments of the project are selected to for study in this paper. One is the new 309m long sewage pipeline, and the other is the new 264.7m long rain pipeline. The site layout is shown in Figure 1. Soil layers distribution around the two pipelines are shown in Figure 2.

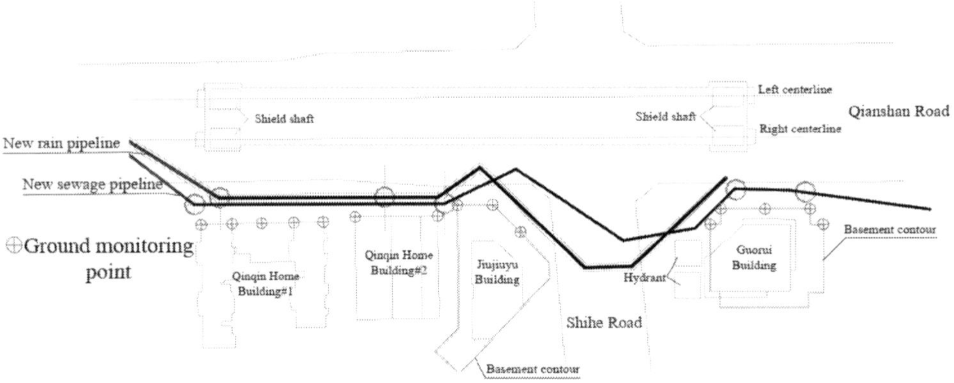

Fig. 1. Site layout

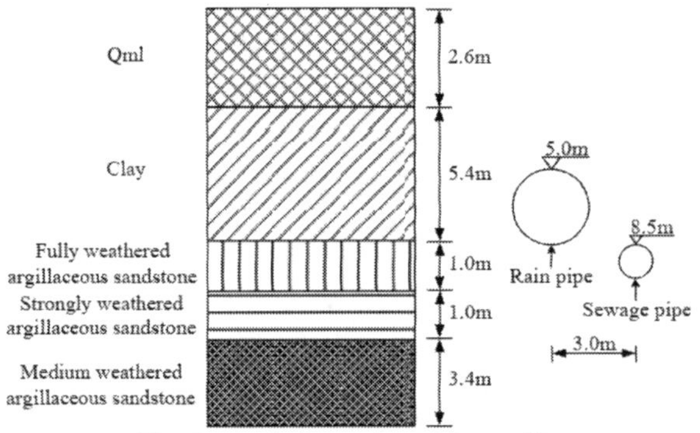

Fig. 2. Typical stratigraphic profile

3 Failure Approach Index (FAI)

FAI, a quantitative index, uses a state variable continuous in space to comprehensively evaluate the dangerous level of rock or soil mass under complex stress state. The calculation formula of FAI is:

$$FAI = \begin{cases} \omega & 0 \leq \omega < 1 \\ 1 + FD & \omega = 1, FD \geq 0 \end{cases} \tag{1}$$

Where $FD = \overline{\gamma}_P / \overline{\gamma}_P^r$, $\overline{\gamma}_P$ is plastic shear strain, $\overline{\gamma}_P^r$ is ultimate plastic shear strain of materials, ω is a phase complementary parameter of YAI, $\omega = 1 - YAI$, YAI is yield approach index. The calculation formula of YAI (detailed deduced process is given in reference 7) under Mohr-Coulomb yield criterion is:

$$YAI = \frac{\frac{I_1 sin\varphi}{3} + \left(cos\theta_\sigma - sin\frac{\theta_\sigma sin\varphi}{\sqrt{3}}\right)\sqrt{J_2} - c\ cos\varphi}{\frac{I_1 sin\varphi}{3} - c\ cos\varphi} \tag{2}$$

where I_1 is the first invariant of strain tensor, J_2 is the second invariant of strain tensor, θ_σ is the so-called Lode's angle, c is cohesion, φ is internal friction angle.

Different stress-strain stages can be defined with FAI. Generally, $0 \leq FAI < 0.8$, $0.8 \leq FAI < 1.0$, $1.0 \leq FAI < 2.0$ and $FAI \geq 2.0$ represent low stress region, excavation disturbance region, excavation damage region and failure region, respectively[8].

4 Numerical simulations

4.1 The geometry and boundary conditions

A three-dimensional numerical model is established in this study based on engineering practice and considering the impact of boundary conditions. The model is 370m in length, 190m in width and 30m in height. It is meshed into eight-node hexahedron elements, as shown in Figure 3. The prescribed displacement boundary conditions are: the bottom surface is fixed, the four sides are set as normal constraint, and the top surface is free in displacement. The building load beside the new pipelines is equivalent to a uniform force of 20kPa per layer and applied on the base of the buildings.

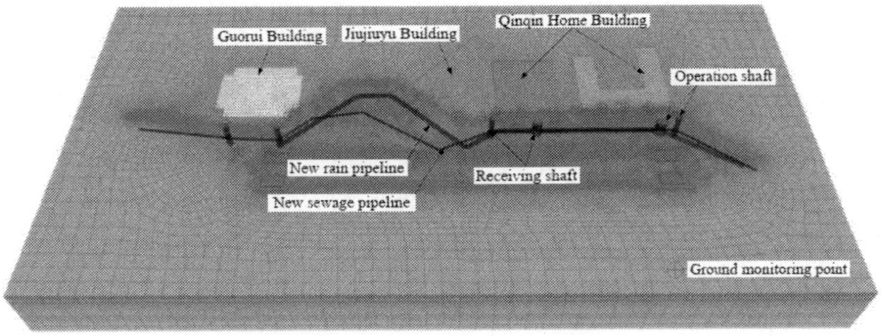

Fig. 3. Layouts of Pipeline and Monitoring Points

4.2 Constitutive model and yield criterion

Normally, in low stress state, the stress-strain curve of rock or soil (including grouting material) shows ideal elastic-plastic characteristic, and the yield criterion of rock or soil accords with Mohr-Coulomb Criterion. Therefore, in the process of numerical simulation, ideal elastic-plastic model is chosen as the constitutive model of rock or soil mass, as well as the grouting material. The new rain and sewage pipe are made of reinforced concrete, and under the action of earth pressure and other external loads, the stress and deformation of reinforced concrete remain in the elastic stage. So the elastic model is chosen as the constitutive model of the pipe.

4.3 Material parameters

According to the engineering geological investigation report and comparing to similar projects, the physical and mechanical parameters of all materials are shown in Table 1.

Table 1. Physical and mechanical parameters of materials

Materials Category	Elasticity modulus (MPa)	Poisson's ratio	Natural unit weight(kN/m^3)	Cohesion (kPa)	Inner friction angle (°)	Tensile strength (kPa)
Plain fill	15	0.20	19.0	15	10	0
Clay	40	0.30	19.8	18	18	0
Fully weathered argillaceous sandstone	105	0.23	18.5	30	20	30
Strongly weathered argillaceous sandstone	110	0.21	20.0	40	30	40
Medium weathered argillaceous Sandston	125	0.20	22.5	50	45	50
Pipe	30000	0.20	25.0	-	-	-
Grout	4	0.30	23.0	15	25	0

4.4 Numerical procedure

Pipe jacking construction is a continuous process. But to facilitate the simulation, it is simplified into a discontinuous step-by-step process. Before the simulation begins, the soil excavation elements, the pipe joint elements and the grouting elements are divided into different groups. It is assumed that the pipe is pushed forward 5m each step in the simulation. Before the jacking begins, all the groups of elements are natural rock and soil mass. In the first jacking step, through changing the material parameters of soil or rock within the first 5m into those of the pipe joint elements and the grouting elements, the first step of the pipe jacking construction can be simulated. Use the same procedure of the first step to simulate the next step of the pipe jacking process until the entire pipe is laid.

5 Analysis of simulation results

5.1 Settlement of the surface near the pipeline

The perturbation caused by pipe jacking construction leads to the settlement of surface above the pipeline[17]. The settlement area is mainly concentrated in the range of about 12.5m from the pipeline, and generally the farther away from the pipeline, the smaller the settlement is. The largest settlement, which is 8.6mm, appears on the surface right above the pipeline, and it is less than 30mm the control value of the ground settlement during construction, as given in Figure 4. The maximum distance between adjacent building and pipeline is about 10m, and the minimum is about 4m. That is to say, all the buildings marked in Figure 4 are influenced by the construction. Furthermore, there are different degrees of settlements on the surface close to buildings as shown in Table 2. Table 2 and Figure 5 show that the simulation results and monitoring data in-situ of the final settlement agree well, which indicates that the numerical simulation process is reasonable and reliable.

The simulation results and monitoring data in-situ show that the settlements of monitoring point JC-29 with the shortest distance to the pipeline is 2.21mm. The maximum settlement 2.83mm happens in monitoring point JC-33 which is 7.15m away from the pipeline. The final settlements of surrounding buildings are less than 3mm, much less than the safety control value 30mm. These show that pipe jacking construction have little effect on surrounding buildings. In fact, because of trenchless construction and the timely follow-up of pipes and grouting, pipe jacking hugely reduces the damage to rock and soil around the pipeline, compared with open cut method, so as to ensure the safety of construction.

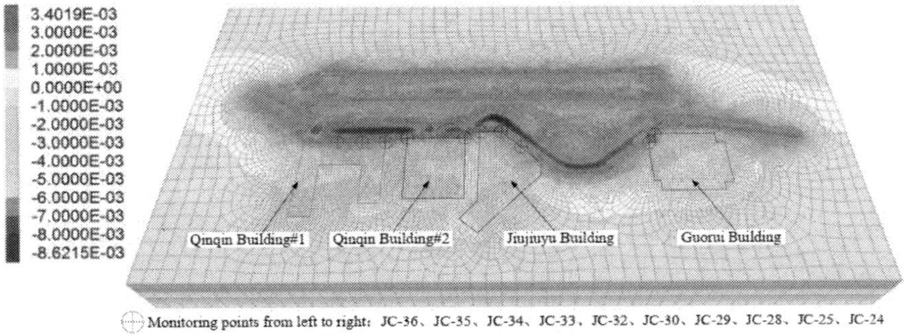

Fig. 4. Overall vertical displacement nephogram

Table 2. Final settlement of monitoring point

Monitoring points	Qinqin Building#1			Qinqin Building#2		Jiujiuyu Building			Guorui Building	
	JC-36	JC-35	JC-34	JC-33	JC-32	JC-30	JC-29	JC-28	JC-25	JC-24
Horizontal distance to pipeline（m）	9.51	9.42	9.41	7.15	6.94	4.12	4.03	4.06	5.22	5.46
Simulation results（mm）	1.17	1.54	1.52	2.83	1.47	1.58	2.21	1.62	1.02	1.37
Field monitoring results（mm）	1.23	1.67	1.63	2.92	1.40	1.60	2.01	1.55	1.12	1.37

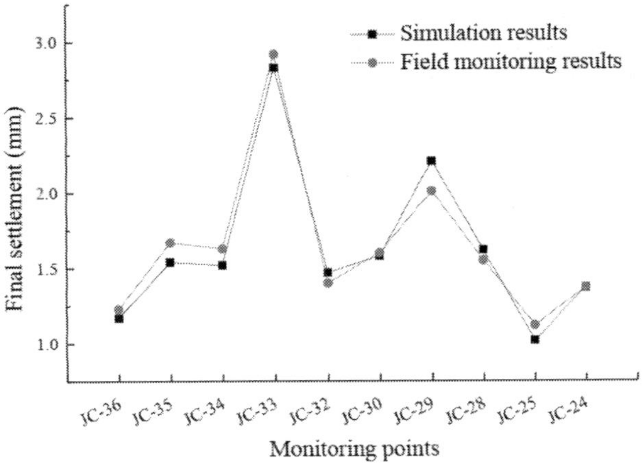

Fig. 5. Vertical displacement of surface monitoring points for building settlement

Taking for example the rain pipeline construction segment between the operation shaft and receiving shaft in Figure 3, making vertical section through the central axis of pipeline, the vertical displacement and the FAI nephogram are shown in Figure 6 and Figure 7. The soil with 0.8≤FAI≤1.2 is distributed within the range of about 0.8 times the pipe diameter above the pipeline, as show in Figure 7. In another word, its most dangerous state still remains in excavation damage region. Near the upper surface of the pipe where the stratum structure disturbance is most severe, the vertical displacement of soil mass reaches

12.7mm. From the positions with 0.8 times of the pipe diameter above the pipeline to the surface, the FAI value of soil mass in this range is around 0.4, that is to say, pipe jacking construction has little influence on the soil mass in this range, due to being in low stress region.

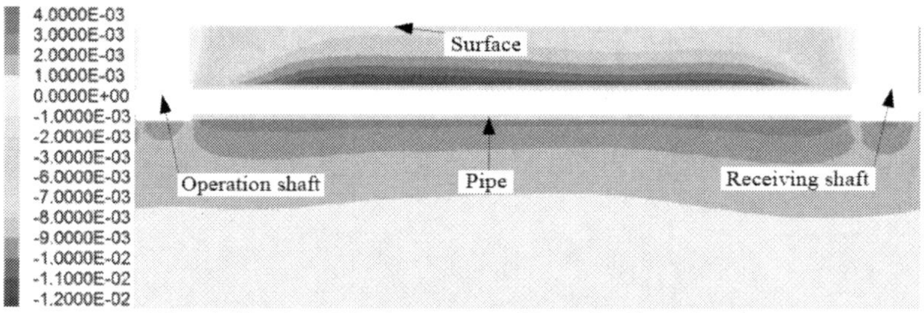

Fig. 6. Vertical Displacement of Longitudinal Section

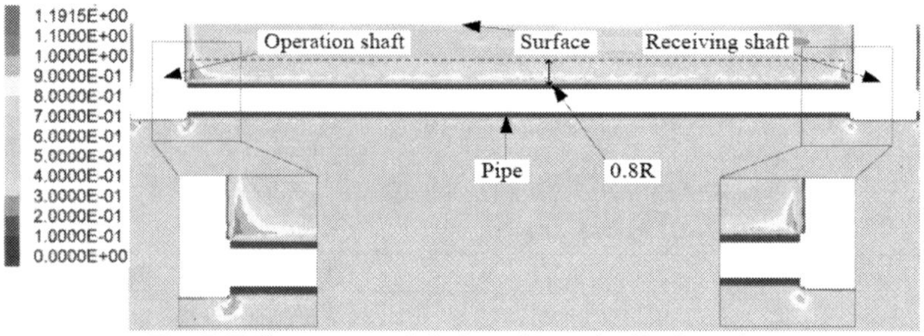

Fig. 7. Failure approach index of Longitudinal Section

5.2 Influence on surface buildings

The influence of construction on surrounding buildings can be evaluated through final settlements only at macroscopic deformation level, yet the final settlements can't reflect the dangerous level and damage degree of rock and soil mass during construction. So FAI is adopted for the evaluation. Figure 9 and 10 are tow cross sections of the pipeline through tow monitoring points JC-33 and JC-29, respectively. Monitoring point JC-33 and JC-29 have the largest final settlement values as shown in table 2. Figure 8 is plotted using the final settlements of a series of monitoring points which are arranged equidistantly on the intersecting lines of the surface and tow cross sections JC-33 and JC-29.

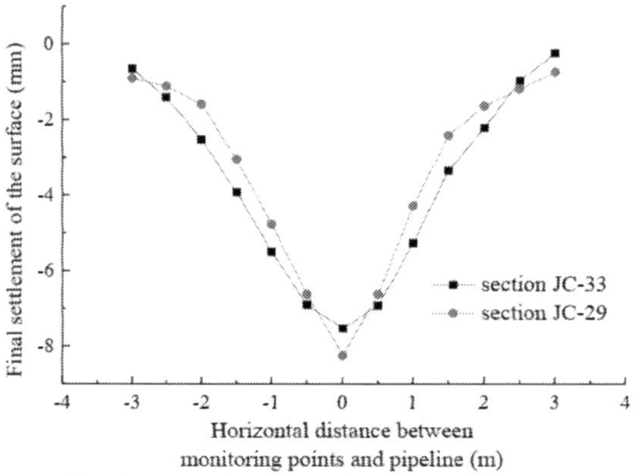

Fig. 8. Subsidence of surface monitoring points perpendicular to pipeline direction

As Figure 9 and 10 showed, distinct deformation area is concentrated in the range of 0.8 times of the pipe diameter near the pipe. And in this area, the rock and soil mass with

0.8≤FAI≤1.3 is in excavation disturbance region and excavation damage region, as shown in Figure 11 and 12. It is to say that the stratum structure in this area is affected by construction more seriously than others. The maximum settlement on surface in this area is 8.6mm. Outside this area, rock and soil mass with FAI≤0.8 is hardly affected by construction, and the maximum settlement on surface is 3.1mm. The building foundations are far away from this area, and there is no obvious change in stress-strain state of the soil near the building foundation. The maximum vertical displacement of the soil near the foundation is about 1mm. Therefore, pipe jacking construction has little influence on surrounding buildings.

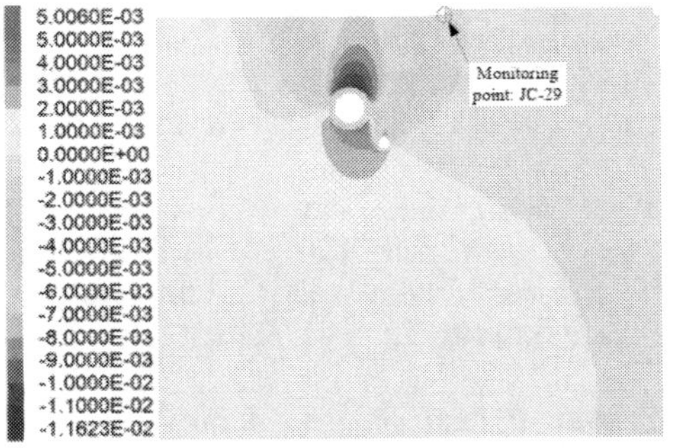

Fig. 9. Vertical displacement of cross section JC-29

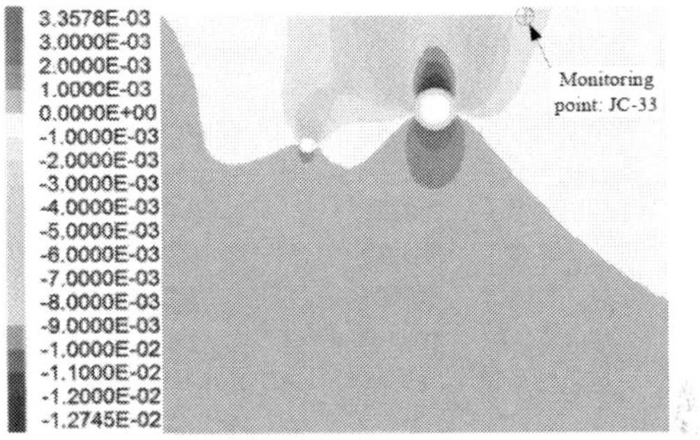

Fig. 10. Vertical displacement of cross section JC-33

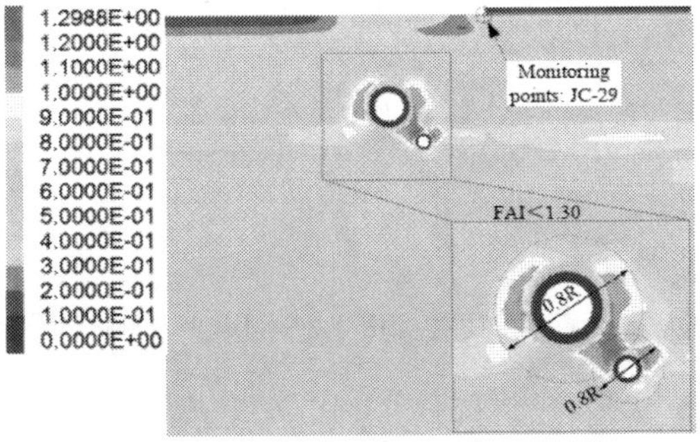

Fig. 11. FAI of cross section JC-29

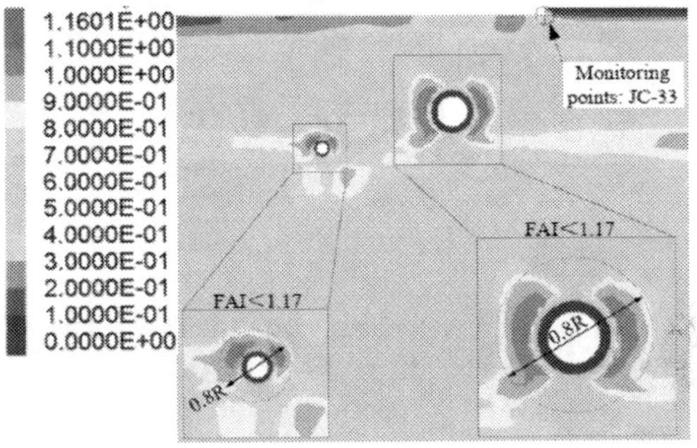

Fig. 12. FAI of cross section JC-33

6 Conclusion

Considering the drawback of plastic zone in traditional simulation methods that it is unable to describe the dangerous level of rock or soil mass, FAI is introduced to evaluate the dangerous state of rock and soil mass during pipe jacking construction, so as to provide scientific basis to the evaluation of safety in construction.

By analyzing the FAI of the rock and soil mass around the pipelines, it turns out that the maximum FAI value is less than 1.3, which proves that there is no sever damage during pipe jacking construction. The range of the excavation disturbance region and excavation damage region (0.8≤FAI≤1.3) is only within 0.8 times of the pipe diameter away from the pipe surface. The stratum structure in this area is disturbed by construction more seriously than others. The maximum settlement on surface corresponding to the area where 0.8≤FAI≤1.3 is 8.6mm, less than the safety control value 30mm. Rock and soil mass near the building foundations with FAI≤0.8 is hardly affected by construction because the building foundations are much farther than 0.8 times of the pipe diameter away from the pipe. The maximum vertical displacement is about 1mm, which is far less than the safety control value. Therefore, pipe jacking is a suitable construction method to be adopted to ensure the safety of surrounding foundations and buildings for the pipe laying.

References

1. R. L. Sterling. Developments and research directions in pipe jacking and microtunneling. Undergr. Sp., (2018)
2. A. I. Sofianos, P. Loukas, C. Chantzakos. Pipe jacking a sewer under Athens. Tunn. Undergr. Sp. Tech., **19**(2):193-203 (2004)
3. H. Feng, X. Gong, R. Xu. Finite element analysis of influence of pipe-jacking construction on environments. Chin. J. Rock Mech. Eng., **23**(7): 1158-1162 (2004)
4. G. Wei, Z. Y. Huang, R. Q. Xu, et al. Study on calculation methods of ground deformation induced by pipe jacking construction. Chin. J. Rock Mech. Eng., **24**(s2): 5808-5815 (2005)
5. F. Li, K. Fang, H. Li. Application of ANSYS 3D FEM in studies of surface deformation caused by pipe jacking. Wuhan Univ. J. Nat. Sci., **12**(4):633-637 (2007)
6. C. Li, Z. Zhong, C. Bie, et al. Field performance of large section concrete pipes cracking during jacking in Chongqing-a case study. Tunn. Undergr. Sp. Tech., **82**:568-583 (2018)
7. Y. Shao, S. C. Wang, Y. Guan. Numerical simulation of soil squeezing effects of a jacked pipe in soft foundation soil and in foundation soil with an underlying gravel layer. Geotech. Geol. Eng., **34**(2):493-499 (2016)
8. C. Li, Z. Zhong, X. Liu, et al. The investigation of ultra-long-distance concrete pipe stuck in quartz sandstone formation using numerical simulation. Arab. J. Geosci., **11**:678 (2018)
9. G. Wei, C. L. Chen, J. Y. Yu. Study on calculating methods of soil vertical deformation induced by pipe jacking construction. Rock Soil Mech., **28** (3): 619-624 (2007)
10. Y. G. Fang, H. Mo, C. Zhang. Theoretic and testing analysis of soil deformation in the area of disturbance caused by pipe-jacking. Chin. J. Rock Mech. Eng., **22**(4):601-605 (2003)
11. H. M. Zhan. Research on ground deformation caused by pipe jacking construction. Chin. J. Undergr. Sp. Eng., **4**(3): 583-585 (2008)
12. F. J. Bing, X. Wang, N. Xi, et al. 3D numerical simulation of pipe jacking and its soil applicability study. Chin. J. Undergr. Sp. Eng., **7** (6):1209-1215 (2011)
13. C. Zhang. Study on method of safety evaluation for rock engineering based on failure approach index. [PhD thesis]. Chinese Academy of Sciences, Wuhan (2006)
14. C. Q. Zhang, H. Zhou, X. T. Feng. Stability assessment of rockmass engineering based on failure approach index. Rock Soil Mech., **28**(5):888-894 (2007)
15. C. Q. Zhang, H. Zhou, X. T. Feng. An index for estimating the stability of brittle surrounding rock mass: FAI and its engineering application. Rock Mech. Rock Eng., **44**(4):401-414 (2011)
16. H. Yao, X. Shao, Z. Zhang, et al. Hydro-mechanical coupling analysis of shield tunnel based on failure approach index. Chin. J. App. Mech., **33**(6):1057-1063 (2016)
17. J. F. Yu. Study on the deformation of earth's surface induced by pipe jacking construction. [PhD thesis], Guangzhou University, Guangzhou (2006)

Applicability of UH model to coarse-grained soil

Xiaolong Zhao[1,2], *Jungao* Zhu[1,*], *Hanbing* Bian[3]

[1]Key Laboratory of Ministry of Education for Geomechanics and Embankment Engineering, Hohai University, 210098 Nanjing, China
[2]Laboratoire de Mécanique, Multiphysique, Multiéchelle, Université de Lille, 5900 Lille, France
[3]Laboratoire de Génie Civil et géo-Environnement, Université de Lille, 5900 Lille, France

Abstract. Large-scale triaxial, K0 compression and constant stress ratio path tests were carried out on two kinds of coarse-grained soils. Stress-strain relationships of these tests were simulated by using UH model for sands to study the applicability of the model to different stress paths of coarse-grained soil, and the results were compared with those of Duncan's E-B model. It is shown that the mechanical behaviors of coarse-grained soil can be well described by the UH model, and overall its capacity to reflect stress-strain relationship under different stress paths is better than that of Duncan's E-B model. Finally, the UH model and Duncan's E-B model were used for stress-deformation analysis of a core rockfill dam. The applicability of UH model for sands to coarse-grained soil was verified.

1.Introduction

With the challenges of climate change and for reducing the emission of greenhouse gas, the renewable energy, such as the hydropower, has been rapidly developed in China [1]. Numerous high earth rockfill dams have been or under constructions in western China. The safety of the hydraulic centre, as well as the dams, the residents and the infrastructures in the downstream area is of major concerns for the hydropower exploitation. A good understanding of the mechanical and/or hydro-mechanical behaviour of the dam is necessary. In fact, for the rockfill dams, the most volumetric part is composed of the so-called rockfill, with an average diameter larger than that we frequently encountered in the laboratories. The classic laboratory test equipment is not suitable for this type of soil. So, the rockfill material is usually scaled into smaller sized material called coarse-grained soil for laboratory test. To avoid the error caused by laboratory test, some in-situ measurements are also developed, but the in-situ measurements have also some difficulties [2-3]. Nowadays, the model parameters used for analysis are still mainly gained by laboratory test [4]. In this context, a series of research on the coarse-grained soil was carried out, with the laboratory test and the numerical approaches. Based on special laboratory tests, a constitutive model has been verified for this type of soil, and then the model has been applied to a rockfill dam. The current article is organised as following, after a short introduction, the model is briefly introduced, then the

* Corresponding author: zhujungao@hhu.edu.cn

experimental research is presented, an application on a rockfill dam is given for the illustration. The discussions and conclusions are given at the end.

At present, many soil constitutive models have been put forward, including nonlinear elastic models [5-8] and elastoplastic models [9-13]. However, most of the models were proposed for clay or sand, and few constitutive models for coarse-grained soil.

Yao et al. [14-15] proposed the three-dimensional unified hardening model for overconsolidated clays (the UH model), based on the relationship between the current yield surface and reference yield surface. The UH model can model many characteristics of overconsolidated clays well, including stress-strain relationships, shear dilatancy, strain-hardening and softening, and stress path dependence behaviour.

In order to describe the similarities and differences between clays and sands in a unified constitutive model, Yao et al. [16] developed the UH model for clays and sands, called UH model for sands.

2. UH model for sands

The UH model for sands, developed by Yao et al. [16], is a constitutive model based on the unified hardening (UH) model [15].

2.1 Shear yielding of sands and its description

A new parameter is introduced into the yield function of the modified Cam-clay model, and is expressed as

$$f = \frac{(1+\chi)q^2}{M^2p^2-\chi q^2} + 1 - \frac{p_x}{p} = 0 \tag{1}$$

where χ is defined as the critical state parameter since it can adjust the vertical distance between NCL and CSL, and it is non-negative and smaller than 1; M is critical state stress ratio; p is mean effective stress, kPa; q is deviatoric stress, kPa; and p_x is intersection of the current yield surface with the p-axis, kPa.

2.2 Dilatancy of sands and its description

The state when the plastic deformation turns from contraction to dilatancy, is a characteristic state used to describe the dilatancy of sands. Here, the equation of the phase transformation line is adopted as that of the characteristic state stress ratio M_c.

$$M_c = M \cdot \exp\left(-m \cdot \xi\right) \tag{2}$$

where m is a dilatancy parameter; ξ is state variable describing the current density.

3. The large-scale triaxial test and K0 test

3.1 Test material and test method

Soils used for tests are one sandy pebble and one gravel. The sandy pebble (denoted by LS) was taken from Dashixia Concrete Faced Sand-gravel Dam, China, while the gravel (denoted by SS) was taken from Maji Rockfill Dam, China. Grain gradations of the two soils are shown in Table 1.

Table 1 Particle size distribution of soil samples

Material	No.	Mass percentage (%) of particle groups (mm)				
		60~40	40~20	20~10	10~5	<5
Sandy pebble	LS	17.6	23.8	18.4	11.7	28.5
Gravel	SS	16.9	22.9	19.6	16.1	24.5

Large-scale consolidated-drained (CD) triaxial tests were conducted on soils of LS and SS with specimen densities 2.04 and 2.20 g/cm^3. The void ratios of LS and of SS are 0.34 and 0.23. The specimen size is 300 mm in diameter and 600 mm in height, and the confining pressures in tests for LS are 600, 1200, 1800 and 2400 kPa, and those for SS are 400, 800, 1600 and 2400 kPa, respectively. The tests were conducted by a large-scale triaxial apparatus, which was in the key laboratory of ministry of education for geomechanics and embankment engineering, Hohai University (Fig. 1(a)). Besides, the constant stress ratio path tests were also carried out by this apparatus, and the stress ratio (σ_1/σ_3) was kept to 2.5 during loading.

In the CD triaxial test, a negative pressure was maintained in the samples for 15 mins before saturation. Then the samples were saturated by allowing water to pass through the base of the triaxial cell and using a top drainage system for removing air voids. The samples were first subjected to the required consolidation pressure and were then sheared by applying axial loading to achieve a rate of loading of 1 mm/min. The shear procedure was terminated when the axial strain reached 15%. In the constant stress ratio path test, the only difference between the CD test is that the σ_1/σ_3 was kept to 2.5 during shearing. The rate of loading was 1 mm/min, and the initial confining pressure σ_3 was 30 kPa in consolidation phase, which would rise with the increase of in shearing.

(a) Large-scale triaxial apparatus

(b) Large-scale K0 apparatus

Fig. 1. Apparatus for laboratory tests

K0 compression tests were performed on LS soil. The apparatus is a large sized oedometer, newly developed by Hohai University [17] (Fig. 1(b)). The maximum axial force can be 800 kN, with an axial stress up to 6.4 MPa. The soil sample was dry, with 400 mm in diameter and 300 mm in height. Stress control was adopted during the test, with a rate of loading of 10 kPa/min.

3.2 Analysis of laboratory test results

Triaxial test results are shown as scattered points in Fig. 2. The test results were also predicted by UH model for sands (hereinafter called UH model) and Duncan's E-B model (hereinafter called E-B model) with parameters given in Table 2 and 3, which are shown with dashed and solid lines, respectively.

Table 2 Parameters of Duncan's E-B model

Material	R_f	K	n	K_b	m	$\varphi_0/(°)$	$\Delta\varphi/(°)$
LS	0.77	530	0.56	364	0.16	52.27	6.77
SS	0.73	1361	0.36	527	0.29	55.48	9.55

Table 3 Parameters of UH model

Material	M	v	κ	λ	N	Z	χ	m
LS	1.92	0.21	0.07	0.15	1.86	0.57	0.87	0.1
SS	1.69	0.18	0.04	0.07	0.95	0.39	0.72	0.1

Fig. 2(a) shows that LS presents strain hardening and shear contraction. There is a good agreement between the model simulations and test results for the two models. Fig. 2(b) shows that SS is strain softening and shear dilatancy. The UH model is obviously superior to E-B model.

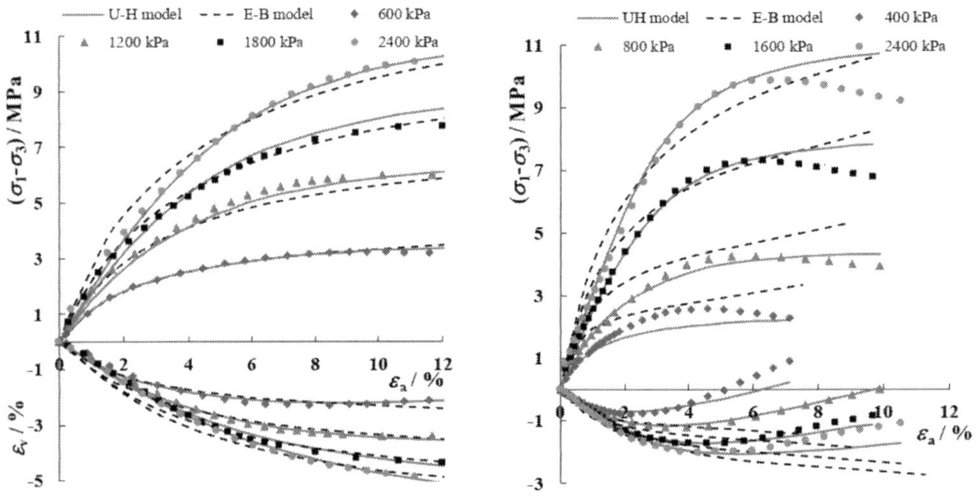

(a) LS (b) SS

Fig. 2. Comparison between triaxial CD test results and model predictions

Fig. 3 shows the predictions of K0 compression test results on LS by the two models. The model parameters used are the same as the triaxial test, given in Table 2 and 3. Apparently, the predictions of UH model is closer to the test data than that of E-B model,

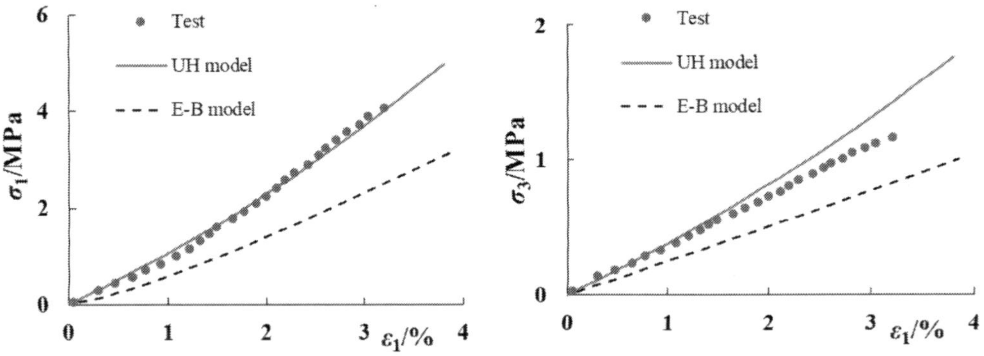

(a) Relationship between σ_1 and ε_1 (b) Relationship between σ_3 and ε_1

Fig. 3. Comparison between K0 compression test results and model predictions on LS sample

Fig. 4 gives the predictions of the constant stress ratio path test results on SS, with the same model parameters (given in Table 2 and 3). The predictions of UH model agrees well with the test results, and there is an evident deviation between the prediction of E-B model and test data for the p-ε_v relationship. For the q-ε_a relationship, both the two models cannot well predict the test results, but the predictions of UH model is still better.

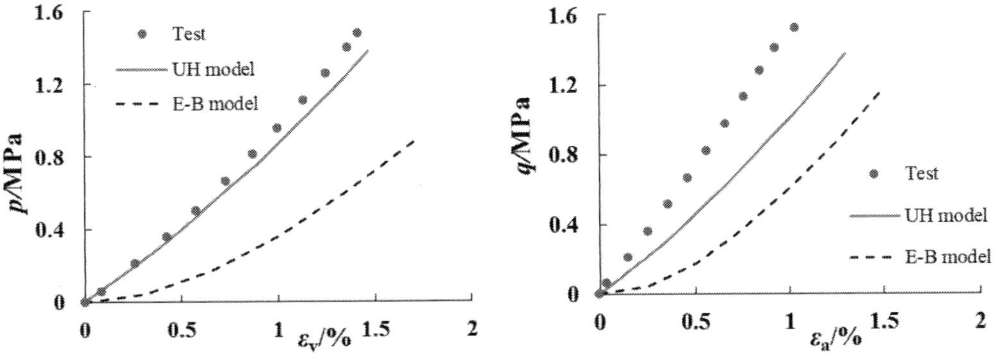

(a) Relationship between p and ε_v (b) Relationship between q and ε_a

Fig. 4. Comparison between constant stress ratio test results and model predictions on SS sample

4. Application of stress-deformation analysis in Rockfill dam

In this section, the stress-deformation analysis of a core rockfill dam was performed by UH model and E-B model, the calculation results were then compared.

4.1 Overview of the dam and finite element model

The earth rockfill dam is 240 m high and the main material zones of the dam are rockfill, transition layer, inverted filter, clay core, ground soil, etc (Fig. 5).

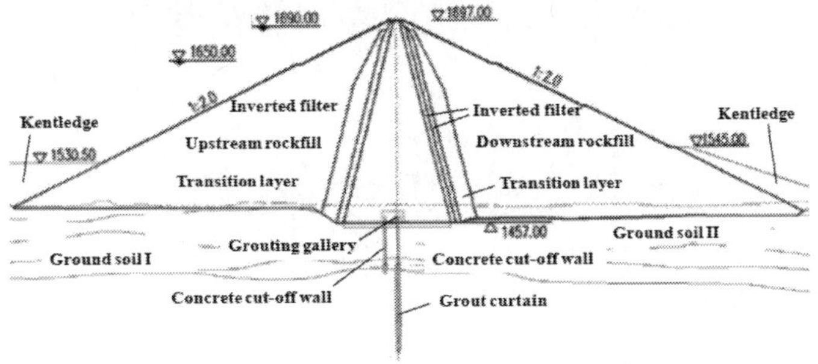

Fig. 5. Typical cross section of core rockfill dam

The model parameters for UH model and E-B model are listed in Table 4 and 5, which were determined with laboratory triaxial test. The concrete material part, such as cut-off wall, was assumed to be linear elastic, with elastic modulus E=30 GPa and Poisson ratio ν=0.17.

Table 4 Parameters of Duncan's E-B model

Material	ρ /(g·cm^{-3})	φ_0 /(°)	$\Delta\varphi$ /(°)	c /kPa	R_f	K	n	K_b	m
Rockfill	2.18	52.9	8.6	0	0.83	1155	0.49	456	0.25
Transition layer	2.25	52.2	7.3	0	0.81	1017	0.32	353	0.21
Inverted filter	2.17	51.2	7.8	0	0.74	1363	0.27	460	0.19
Clay core	2.21	30.4	0	96	0.79	402	0.44	204	0.42
Ground soil I	1.45	48.0	7	0	0.80	1100	0.34	360	0.32
Ground soil II	1.31	46.9	6.5	0	0.77	1075	0.33	343	0.27

Table 5 Parameters of UH model

Material	M	v	κ	λ	N	Z	χ	m
Rockfill	1.700	0.11	0.05	0.169	1.887	0.567	0.28	0.25
Transition layer	1.699	0.20	0.05	0.192	1.900	0.272	0.43	1.37
Inverted filter	1.700	0.19	0.02	0.100	1.155	0.462	0.23	0.11
Clay core	1.487	0.40	0.01	0.100	1.170	0.593	0.44	0.55

There are 17780 nodes and 19363 elements in the finite element model. Most of the elements are 8-node hexahedrons, with a few others, such as 6-node pentahedrons and 4-node tetrahedrons for transition. Fig. 6 shows the three-dimensional finite element mesh of the dam.

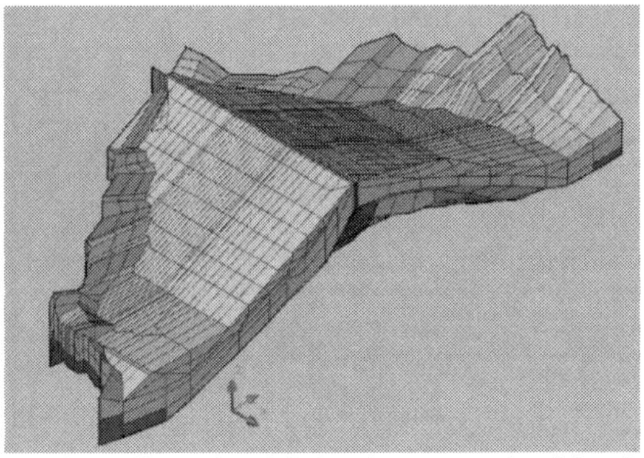

Fig. 6. Three-dimensional finite element mesh of dam body

4.2 Calculation results comparison between models

The settlement of rockfill dam is usually about 1%~2% of the dam height, and displacement ratio (maximum horizontal displacement along river to maximum settlement) is less than 0.25 [18]. In this paper, the settlement from E-B model accounts for 1.3 % of the dam height (including the ground soil, 50 m), with the displacement ratio 0.18, while the settlement ratio from UH model is 0.9%, with the displacement ratio 0.19, which are all within the reasonable range.

Fig. 7 shows the contour lines of major principal stress in the maximum cross section. Due to the so-called arch effect, the major principal stress in core wall is smaller than that in transition layer at the same elevation. The arch effect calculated by E-B model is stronger than that by UH model.

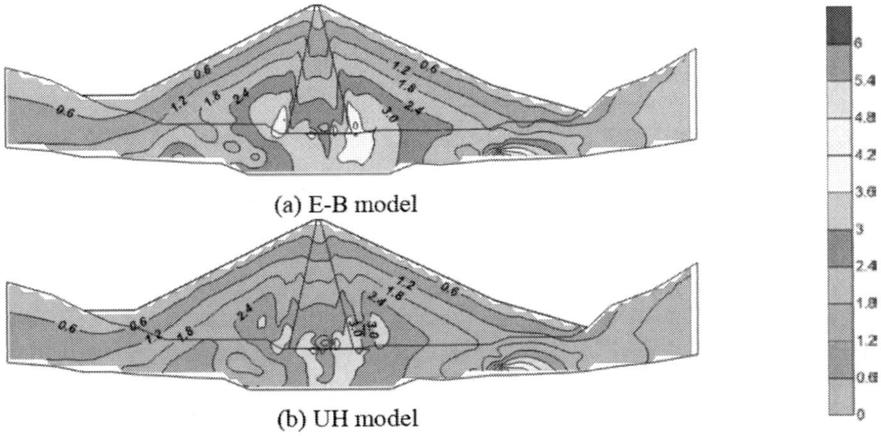

(a) E-B model

(b) UH model

Fig. 7. Contour lines of major principal stress at completion (unit: MPa)

5. Discussion and Conclusions

In this paper, large-scale triaxial test and K0 compression test were conducted on two coarse-grained soils. The test results were predicted by UH model and by E-B model. By comparison, the UH model was better than E-B model in the prediction of these test data. Furthermore, the UH model can be well applied in the stress-deformation analysis of rockfill dam, as shown in section 4. The following conclusions are drawn:

(1) The stress-strain relationship of triaxial CD test can be relatively well described by both UH model and E-B model. However, for the coarse-grained soil with strain softening and shear dilatancy, E-B model has a poor ability of simulation, and it cannot reflect these behaviours while UH model can describe these properly.

(2) Simulations of K0 compression test and constant stress ratio test show that UH model can well predict stress-strain relations under complex stress paths, and the predictions are better than that of E-B model.

(3) Finite element analysis indicates that UH model can be applied in the calculation of earth rockfill dam.

Acknowledgements

The authors gratefully acknowledge the financial support from the National Key Research and Development Program of China (Grant No. 2017YFC0404801), the National Natural Science Foundation of China (Grant No. 51479052) and the Chinese Scholarship Council.

References

1. Z.Z. Yin. Stress and deformation of high earth and rock-fill dams. Chinese Journal of Geotechnical Engineering, **31** (1), 1-14 (2009) (in Chinese).
2. W.J. Xu, Q. Xu, R.L. Hu. Study on the shear strength of soil–rock mixture by large

scale direct shear test. International Journal of Rock Mechanics and Mining Sciences, **48** (8), 1235-1247 (2011).

3. M. Sanei, L. Faramarzi, A. Fahimifar, S. Goli, A. Mehinrad, A. Rahmati. Shear strength of discontinuities in sedimentary rock masses based on direct shear tests. International Journal of Rock Mechanics and Mining Sciences, **75**, 119-131 (2015).
4. G.X. Li. Some problems in researches on constitutive model of soil. Chinese Journal of Geotechnical Engineering, **31** (10): 1636-1641 (2009) (in Chinese).
5. T.L. Konder. Hyperbolic stress-strain response: cohesive soils. Journal of Soil Mechanics and Foundation, **89** (1), 115-143 (1963) (ASCE).
6. J.M. Duncan, C.Y. Chang. Nonlinear Analysis of stress and strain in soils. Journal of the Geotechnical Engineering Division, **96** (5), 1629-1653 (1970) (ASCE).
7. J.M. Duncan, P. Byrne, K.S. Wong, P. Mabry. Strength, stress-strain and bulk modulus parameters for finite element analysis of stress and movement in soil masses. Berkeley: University of California (1980).
8. Z.L. Cheng, J.S. Jiang, H.S. Ding, Y.Z. Zuo. Nonlinear dilatancy model for coarse-grained soils. Chinese Journal of Geotechnical Engineering, **32** (3), 460-467 (2010) (in Chinese).
9. K.H. Roscoe, A.N. Schofield, A. Thurairajah. Yielding of clays in states wetter than critical. Geotechnique, **8** (2), 22-53 (1963).
10. K.H. Roscoe, J.B. Burland. On the generalized stress-strain behaviour of wet clay. Cambridge University Press, Cambridge, 535-609 (1968).
11. P.V. Lade, J.M. Duncan. Elastoplastic stress-strain theory for cohesionless soil. Journal of the Geotechnical Engineering Division, **101** (10), 1037-1053 (1975).
12. K.M. Wei, S.S. Chen, G.Y. Li, Z.Z. Fu. Constitutive model for coarse-grained dam materials considering state parameter. Chinese Journal of Geotechnical Engineering, **38** (4), 654-661 (2016) (in Chinese).
13. S.H. Bian, G.Y. Li, K.M. Wei, J. Zhou. Modified generalized plasticity model for rockfill materials. Chinese Journal of Geotechnical Engineering, **39** (10), 1936-1942 (2017) (in Chinese).
14. Y.P. Yao, W. Hou, A.N. Zhou. Constitutive model for overconsolidated clays. Science in China Series E: Technological Sciences, **51** (2), 179-191 (2008).
15. Y.P. Yao, W. Hou, A.N. Zhou. UH model: three-dimensional unified hardening model for overconsolidated clays. Geotechnique, **59** (5), 451-469 (2009).
16. Y.P. Yao, L. Liu, T. Luo, Y. Tian, J. M. Zhang. Unified hardening (UH) model for clays and sands. Computers and Geotechnics, **110**, 326-343 (2019).
17. J.G. Zhu, M.J. Jiang, Y.Y. Lu, W.L. Guo, B. Zhang. Experimental study on the K_0 coefficient of sandy gravel under different loading conditions. Granular Matter, **20** (3), 1-9 (2018).
18. X.L. Zhao, J. G. Zhu, P. Wang. A comparison of two constitutive models in stress and deformation analysis of earth rockfill dams. China Rural Water and Hydropower, 1, 165-169, 173 (2018) (in Chinese).

Synthesis of numerical methods for the design of segmental tunnel lining

Rim Trad[1,*], *Hussein* Mroueh[1] , *Hanbing* Bian[1] ,*Fabrice* Cormery[2]

[1] Laboratoire de Génie Civil et géo-Environnement, Université de Lille, 59655 Villeneuve d'Ascq, France

[2]LAMCUBE, Université de Lille, 59655 Villeneuve d'Ascq, France

Abstract. This paper presents a numerical study that aims to compare the behavior of the segmental tunnel lining using the direct, indirect and experimental methods. This model is based on a practical case applied in university of Tongji: a project of water conveyance tunnel. A reduction in the bending moment and increasing of the displacement in the tunnel lining is showed in numerical results, when taking into account the effect of the joints. It has been shown that the number of joints in the tunnel-lining structure highly affects the results in terms internal forces and displacements. Furthermore, the internal forces obtained by the continuous method are high compared to the other methods when the effects on segmental joints on tunnel lining behaviour are usually considered. Additionally, the bending moment of the direct method with behaviour of rotation spring linear and experimental method is comparable.

1 Introduction

The mechanized shield tunnelling method has become more and more popular for the construction of tunnels in recent periods due to its flexibility, cost effectiveness and minor trouble to the surrounding ground [19]. Because of the presence of joints between segmental linings driven in place by tunnelling operations, its design should take into account these discontinuities. Generally, the influence of joints on the internal forces and displacement is one of the main issues in the segmental lining design of shield-driven tunnels.

In order to describe the influence of joints in tunnels consisting of segments, various methods have been investigated in the literature, from theoretical to numerical (two-dimensional or three-dimensional) models, and compared to experimental data's with some accuracy (references).

Recently, a research program at Tongji University [22] has been conducted to study the three-dimensional effect through full-scale loading tests on a staggered-fabricated three-ring structure for rain-water storage and transportation tunnel. Based on their preliminary results, we propose in this paper a comparative numerical analysis of segmental lining modelled as uniform or discontinued segments, either by continuous, indirect and direct

* Corresponding author: rimtrad_7@hotmail.com

methods. In this model, in addition from the soil pressures and the normal external water, inner hydrostatic pressures will also applied during strong storms. Single-ring loading tests have been realized to evaluate under the designed loads the performance of the segmental lining. Results are compared with experimental results obtained from the project of Tongji University [22].

2 Survey of existing methods

There are two main analytical methods for taking into account the effects of joints between segments on tunnel behaviour: indirect and direct methods.

In indirect method, tunnel structure is perceived as a rigid lining ring embedded on a continuous ground model [1], [2],[3],[4],[6],[11]. The presence of joints causing a decrease in the rigidity of the structure, these methods simply consist in considering all the segments as a continuous beam ring with a reduced rigidity, i.e. by applying a reduction factor to the rigidity in bending of the elements of the tunnel. These simplified analytical methods cannot take into consideration the intricacy of the joint characteristics, including joint distribution and joint stiffness, nor analyze complex situations of the surrounding ground.

In direct method, the segments are modelled by beam elements linked together by a connection with rotation spring supposed to describe the joint [7], [9], [10], [12]. The behavior of the springs in rotation can be linear or nonlinear. [10] Proposed a numerical method where joint behaviour is assumed by means of all three joint stiffness's, that is, the rotational stiffness, the axial stiffness and the radial stiffness. On the other hand, these methods cannot be applied to cases in which the joint distribution is asymmetrical to the vertical axis of the tunnel [16].

To better understand the local behavior within the structures formed of segments, many experimental tests were conducted. They made it possible to highlight the role played by the joints and the technologies used [21], [20], [14]. For example, [5] performed various experimental tests on flat joints with a concrete-concrete interface (with and without bolts) and determined the experimental moment (M) - rotation (θ) relationship. Others like [17] have studied the flat bolted longitudinal joint structural behavior of Shanghai Metro Line No.13 with a full-scale test conducted continuously loading the joint until it is completely failure.

For example, In particular, they have shown that the bolt position along the surface contact influence the joint behavior but also its behavior itself. Other researchers [18] conducted a series of large-scale experimental trials on two types of seals used in a water transfer tunnel project. The main objective of this study was to better understand the mechanism of rotational behavior of these two types of joints in cases of negative and positive bending moment. This brief review of the experimental literature on complex structures provides a better understanding of the overall behavior of structures formed by segments and linked together by joints.

Concerning numerical approach, by using numerical tool, the Flac [15] modelled a tunnel for the railway line Bologna-Florence. A 2D model has been developed under condition of plane deformations. It is supposed that the behavior of the tunnel structure is linear elastic and that of the ground is defined by an elastic perfectly plastic constitutive relation, that is based on the Mohr–Coulomb failure criterion. A segmental joint can be simulated as an elastic pin and its stiffness characteristics are affected by rotational stiffness KRO, radial stiffness KR and axial stiffness KA.

Through of simplified FEM 3D analyses, [13] presented a numerical study. They studied the influence of joints (The joint are modelled with interface elements, indicated with springs). They calculated a range of rotational stiffness between 1000 and 3000 kN.m/rad, then a constant elastic rotational stiffness is used to model the joint. The interaction between the soil-structure is also taken into account through a set of normal reaction springs, without taking into account tangential springs. The authors investigate also the influence of longitudinal joins number and position in the transversal lining section.

To better understand the behaviour observed in experiments, [18] developed a three-dimensional finite element (FEM) model using ABAQUS (version 6.14). In particular, they took into account: the numbers of bolts, cast iron panels and initial conditions of contact between the different segments. The presence of gasket that produces a gap between the segments has an important influence on longitudinal joint rotation rigidity. They concluded to the presence of gaskets produces a space between the segments, this space have a significant influence on the longitudinal joint rotation rigidity.

3 Project background

The full-scale loading tests were carried out based on a pilot 'deep tunnel' under Suzhou River of Shanghai [22], which purpose to extenuate the waterlogging and ground overflow problems of this city during the heavy raining seasons. A tunnel length of 1.5 km has testing, and the average burial depth to the tunnel crown is about 50 m. Fig.1 show the cross-sectional layout of the lining structure that has a thickness 0.65 m and an inner diameter of 9 m. A single lining ring has a width of 1.5 m and composed of eight segmental blocks, including one key block (K), two adjacent blocks (AB) and five identical standard blocks (SB). The precast C60 concrete reinforced with steel fiber is adapted to lining structure. All this data is extracted from [22].

Fig.1. Cross-sectional layout of the lining structure

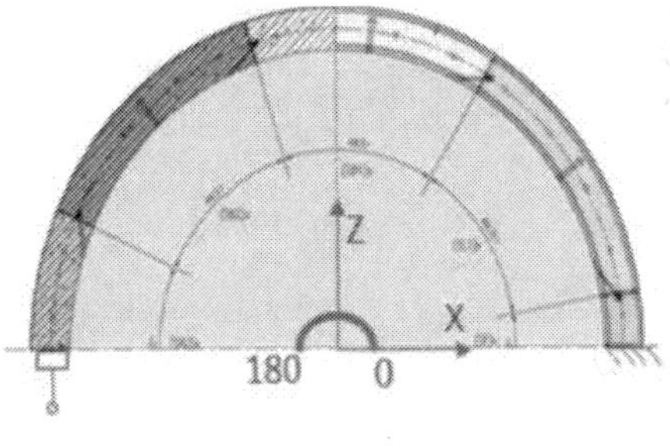

Fig.2. : Half-ring model

In the present study, a symmetrical half-ring model is used to reduce computation time (Fig.2.). In addition, the segments is modeled as a beam, it is fixed at one end and supported at the other (vertical displacement is zero and rotation are blocked).

Fig.3 presents the loads following up on the testing specimen considered in the present investigation extracted from experimental study [22] and the load symbols is given in Table 1.

Table 1. Load symbols

Symbols	Loads
p	Overburden earth pressure
(q1),(q2)	Lateral earth pressure
qw0	External water pressure
qwi	Inner water pressure
k	Coefficient of lateral resistance
δ	Deflection of segmental lining [−45°; 45°]
pk	Subgrade reaction pressure
g	Segment's self-weight

In the experimental installation [22], these mechanical loads are converted into 16 sets of concentrated loads, between then the 15 sets loads are symmetrically dispersed according to the vertical axis of the testing configuration and last one set of loading points uniformly distributed over the periphery of the testing configuration. This last, was used in the three-ring tests to apply the longitudinal thrust forces. But in this study, just the single-ring tests are investigated, so no longitudinal thrust forces were applied. Therefore, 15 sets loads are applied.

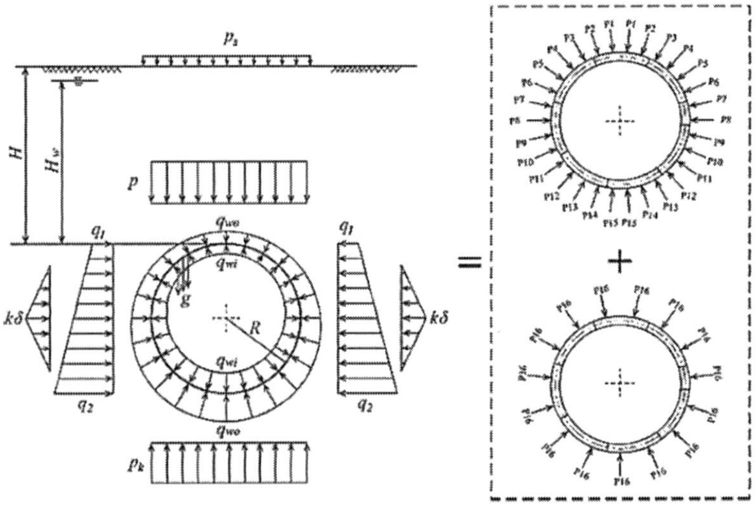

Fig.3. : Loads acting on the testing specimen [22].

4 Research methods

In view of different methods in literature, three methods of segmental linings were designed for this study as shown in. Several methods are investigated in this study: direct and indirect methods. The continuous method assumes the tunnel ring as continuous beam, without presence of joint. Direct and indirect methods are applied to determine the influence of joint in tunnel, by means of a finite element model and compared with an experimental result obtained from project of Tongji University [22].

As long as indirect methods are affected, the effect of joints is generally taken by a reduced rigidity of the tunnel structure; its value is between 0.25 and 1.

In direct methods, segments and joints are modelled as structural elements. Segments and joints are represented as structural beam elements and rotational springs respectively. The behavior of rotating springs is assumed linear with constant rotational rigidity 50000 KN.m/rad.

5 Results

The variation of displacement obtained throughout tunnel (fig.2) for the different methods is shown in fig.4. In this figure, the displacement increases throughout tunnel for indirect, direct and continuous methods. According to the indirect method, it is noted that as the joint stiffness factor increases, the displacement decreases. Because, more the joint stiffness factor, η is close to 1, more the model behaves like a continuous tunnel so has a lower displacement compared to the other method. Moreover, direct model is calibrated through the rotational spring value, to obtain the same maximum displacement that experimental study 8 cm (Fig.4).

The numerical results show a significant reduction of bending moment when the effect of distribution of joints is taken. It is observed that the bending moments obtained by the continuous method are higher than those of the direct and experimental methods where the effects on the joints between the segments are taken into account (Fig.5.). It is found that the bending moments of direct method and experimental method are comparable (Fig.5).

(In the figures, the simplified moments represent the ratio of the moment obtained from the calculation and maximum moment obtained in continuous method).

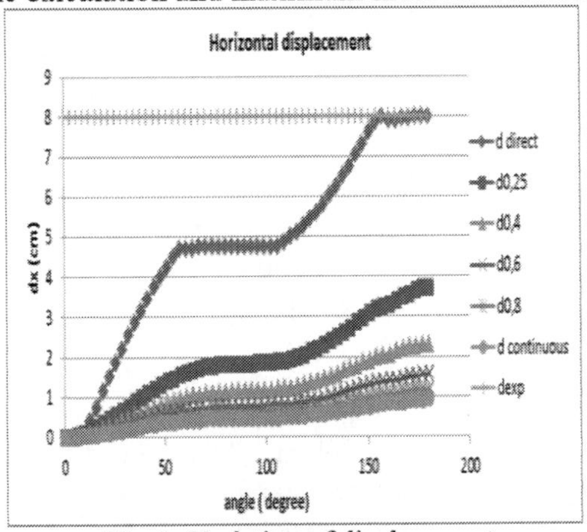

Fig.4.: Variation of displacement

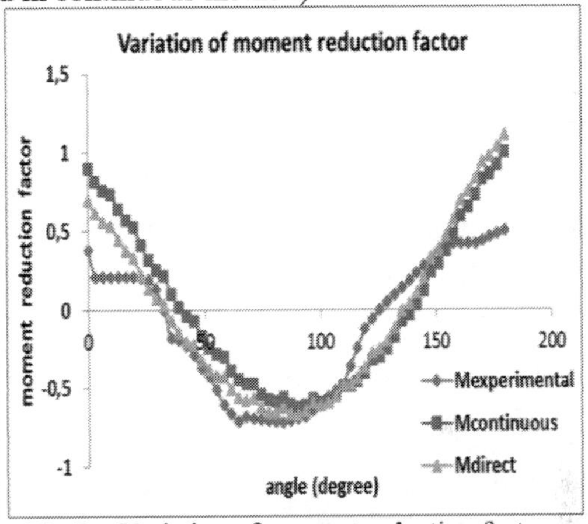

Fig.5. : Variation of moment reduction factor

Instead, it can be seen (Fig.6.) that the number of joints in a tunnel has a great influence on the results in terms of bending moment. Thus, when the number of joints increases, the bending moment increases.

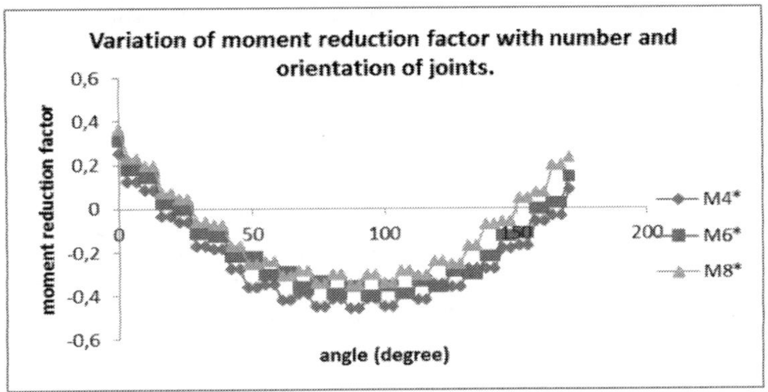

Fig.6. : Variation of moment reduction factor with number and orientation of joints.

The number of joints has an important effect on the displacement as a function of the joint rotational stiffness as illustrated in Figures 7 and 8. When the number of joints increases, the displacement is smaller. It can be noted that the influence of rotational rigidity in the range (4000 to 20000 KN.m / rad) is very important (Fig.7.) for the displacement (Fig.8.). The higher the rotational stiffness joint, the greater the displacement decreases sharply.

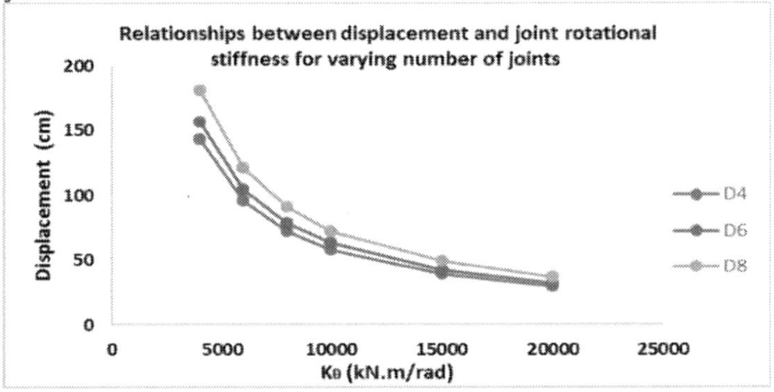

Fig.7. : Relationships between displacement and joint rotational stiffness for varying number of joints

6 Conclusion

In this study, a comparative numerical study for different methods, representing the segmental lining as a uniform or discontinued segment is presented. Different methods are applied on a finite element model and compared with the experimental results got in the project of Tongji University.

Based on the results, the numerical results show a significant reduction in the bending moment and increasing of the displacement in the tunnel lining taking into account the effect of the distribution of joints. Besides, the bending moment obtained by the continuous method are high compared to the other methods when the influence of segmental joints on tunnel lining behaviour is usually considered. Additionally, the bending moment obtained from the direct method (with law behaviour of rotation spring is linear) and experimental method are comparable. Furthermore, it has been shown that the number of joints in the tunnel-lining structure highly affects the results in terms bending moment and displacements.

In view of the results, it seems that taking into account of the rotation is primordial. It is then necessary to describe the joint behaviour in the most relevant way that is depending on the geometry and characteristics of joints. Our work aims to propose a friendly-method that is capable to take into account the complex local behaviour of joints by way of global approach with globalized parameters, without dealing with a complex three-dimensional and non-linear calculation.

Following the limitation of the current numerical methods, we propose to model the joint using the concept of macro-element, which consists in considering all material and geometrical nonlinearities of joints locally and transferring the local behaviour to some global structural parameters between linings, by concentrating them in a representative point and by formulating a three-dimensional law in generalized variables (moment and rotation).

References

1. Wood, A. M. J. Geotechnique, **25**, 115-127(1975).
2. H. H. Einstein, C. W. Schwartz, J. Geotech Eng Div. **105**, ASCE 14541 (1979).
3. H. Duddeck, J. Erdmann. J. Underground Space **9**, 246–259 (1985).
4. J.H. Liu, and X.Y. Hou. J. China Railway Press. 152–303 (1991).
5. Hordijk, D. A., F. B. J, Gijsbers, and Projectbureau Boortunnels. J. Reporte Interno, (1996).
6. No, W. G., & International Tunnelling Association. J. TUNN UNDERGR SP TECH **15**, 303–331 (2000).
7. K. M. Lee, X. Y. Hou, X. W. Ge, Y. Tang. J. INT J NUMER ANAL MET **25**, 365–390 (2001).
8. Blom, C. B. M. J. of the fib **4**, 89–94 (2002).
9. W Ding, Z Yue, L Tham et al. J. INT J NUMER ANAL MET **28**, 57–91 (2004).
10. P. P. Oreste. J. TUNN UNDERGR SP TECH. **22**, 185–205 (2007).
11. H. El Naggar, S. D. Hinchberger, .J. Can. Geotech**45**, 1572–1593 (2008).
12. S. Teachavorasinskun, T. Chub-uppakarn, .J. TUNN UNDERGR SP TECH **25**, 490–494 (2010).
13. S. H. P. Cavalaro, C. B. M. Blom, J. C. Walraven, A. Aguado, .J. TUNN UNDERGR SP TECH **26**, 734–749 (2011).
14. N. A. Do, D. Dias, P. Oreste, I. Djeran-Maigre. J. TUNN UNDERGR SP TECH **37**, 115–127 (2013).
15. N. A. Do, D. Dias, P. Oreste, I. Djeran-Maigre. J. INT J NUMER ANAL MET **38**, 1617–1632 (2014).
16. X. Li, Z. Yan, Z. Wang, H. Zhu. J. ENG STRUCT **93**, 97–113 (2015).
17. Y. Jin, W. Ding, Z. Yan, K. Soga, Z. Li. J. TUNN UNDERGR SP TECH **68**, 153–166 (2017).
18. X. Liu, Z. Dong, Y. Bai, Y. Zhu. J. TUNN UNDERGR SP TECH **66**, 1–18 (2017).
19. L. Zuo et al. J. TUNN UNDERGR SP TECH **77**, 227–236 (2018).
20. M. Karami,S. Zare. J. CEEJ, 3-18 (2018).
21. X. Huang et al. J. TUNN UNDERGR SP TECH **88**, 156–168 (2019).

AUTHOR INDEX